Toward a More
Sustainable
Agriculture

Toward a More
Sustainable
Agriculture

Raymond P. Poincelot
Biology Department
Fairfield University
Fairfield, Connecticut

AVI Publishing Company, Inc. Westport, Connecticut

Copyright © 1986 by Raymond P. Poincelot

Published by AVI Publishing Company, Inc.
          250 Post Road East
          P.O. Box 831
          Westport, Connecticut 06881

Library of Congress Cataloging-in-Publication Data

Poincelot, Raymond P., 1944–
    Toward a more sustainable agriculture.

    Bibliography: p.
    Includes index.
      1. Agricultural conservation—United States.
    2. Agriculture—United States—Energy conservation.
    3. Agriculture—United States.   4. Organic farming—
    United States.   I. Title.
    S604.6.P65   1986   630   85–26786
    ISBN 0–87055–518–9

Printed in the United States of America

ABCDE 5432109876

To my loving and supportive family:

My wife Marian and our children,
Raymond, Daniel, and Wendy

May the food and fiber system
endure for their children
and subsequent generations

Contents

Contents (continued)

Contents (continued)

Preface

Our nation's grandest enterprise is our agricultural industry. It is second to none in terms of assets, workers, and exports. Agricultural success has become an accepted fact and is taken for granted by the majority of the American public. Few believe or are even willing to consider that the continued future success of this industry is threatened.

Yet threatened it is. The resource base of agriculture is becoming diminished through overuse and environmental misuse. A further complication is the competition for agricultural resources by other users. The energy, soil, and water resources cannot sustain agriculture into the far future at their present rate of use.

Something must be done to bring about public awareness and support for the changes needed to move our nation toward a sustainable agriculture. More research and funding must be directed toward this end. Our agriculture educators and other information disseminators must make sure that the farmers, politicians, and the public receive the message. Farmers must be willing to make the necessary changes.

Something is being done. Our agricultural system is in a transitional stage. Traditional agriculturists are changing some practices and their attitudes. Alternative agricultures have appeared. The foot has been set on the start of the path. The path is long, and may have false forks and dead ends along the way. Hopefully for our future generations we shall persevere. The way undoubtedly will be difficult, even painful, and certainly costly. We have no choice, however, if we wish to remain as a viable society.

For these reasons this book was written. Through examination comes discussion and ultimately action. In the hopes of reaching many, this book was written for several audiences in the agricultural and nonagricultural communities. These groups include, but are not restricted to, the educators, extension workers, farmers, food and fiber processors, researchers and students in the agricultural community, and those beyond who share an interest in and a concern for agriculture. The latter include civic-minded citizens, planners and policy makers, politicians, and those involved in environmental or ecological activities.

As such, the format is variable and perhaps not all information is relevant to any one group. This wide approach out of necessity involves reviews, examination of practices and technologies, and future projections. Sometimes the material is simple, and other times highly technical. The progression is an examination of energy use throughout the food and fiber system, soil and water resource overuse, environmental problems, and future technologies.

In closing I would like to thank J.R. Ice and W.W. Tressler of AVI Publishing Company for their cooperation, efforts, and support toward the publishing of this book. Also, it should be noted that prior to my present affiliation, I was an Associate Agricultural Scientist with The Connecticut Agricultural Experiment Station in New Haven, Connecticut.

I would welcome your comments and criticisms at my present address: Dean of Education, New York Botanical Garden, Bronx, New York 10458-5126.

Our Threatened Agricultural Resources

Agricultural production in the United States has almost become synonymous with success. Certainly, no one would argue against its being a cornerstone of American greatness, nor suggest that the significant yield increases and corresponding labor decreases of the past several decades were undesirable. Certainly not. Yet, an uneasiness has begun to surface in certain quarters, as the long-term costs of American agriculture gradually become apparent.

This concern is not yet widespread, since its warning message is lost in the clamor about agricultural success. After all, agriculture is the biggest U.S. industry (USDA 1983) with assets presently over one trillion dollars. Such a staggering sum is about 70% of the capital assets of all U.S. manufacturing corporations. Such bigness is also reflected in national employment figures: Agriculture is the nation's largest employer, providing one of every five jobs in private enterprise. Some 23 million individuals are employed in agricultural enterprises associated with all phases of agriculture from production through food sales. The actual work of food production—farming—employs some 3.4 million workers. This number is about equal to all of those employed in transportation and the steel and automobile industries. The remaining 20 million provide farming supplies and are involved in the postproduction aspects of food and fiber: storage, processing, transportation, and merchandising.

All the remarkable production of U.S. agriculture comes from about 2.4 million farms. One farmworker supplies food and fiber for 78 people. Today 1 hr of farm labor produces 14 times what it did in 1919–1921. With the same production base, crop and livestock output has more than doubled over the last 50 years. Such success has led to the good life and made the United States first in the world as an exporter of farm products.

Yet, a cloud does appear on the horizon. It will not go away and it cannot be ignored. The continued success of agriculture in the future will be affected by this gathering storm. The problems of agriculture touch us all. Failures will result in tremendous disruption in our socioeconomic and political systems. We have a warning light in those problems of today, and they demand careful examination.

Agricultural Problems

The main future threat to agriculture is a diminishing resource base. Resources are threatened in two ways: by depletion and by contamination such that the resources become unusable. Both problems have an impact far beyond agriculture, in that the resulting loss of food production and environmental damage threaten and diminish our quality of life. The socioeconomic and political consequences, if current trends are allowed to continue unabated, are grim.

Undoubtedly some knowledgeable farmers, planners, policy makers, and scientists were aware of agriculture's diminishing resource base in the past. In fact, this awareness was responsible in part for the development of a policy on the conservation of natural resources as a component of national agricultural policy formulated in the 1930s. However, this awareness was certainly strengthened and acted upon because of a highly visible, dramatic event: the Dust Bowl of the 1930s.

Concern about the depletion and contamination of agricultural resources also probably was a motivating factor in the rise of the organic gardening and farming movement in the 1940s, and the subsequent development of commercial organic farms over the next three decades. This awareness also played some role in the creation of various environmental groups over this same period, eventually culminating in the 1970 creation of the federal Environmental Protection Agency.

However, it was not until the 1970s that awareness of agricultural resource problems began to enter the mainstream of national consciousness. Several events contributed to this awakening. An early milestone was the appearance of *The Limits to Growth* (Meadows *et al.* 1972, 1974), which carried a somber message on the finiteness of resources, including agricultural resources. The environmental impact of pesticides resulted in passage of the 1972 Federal Environmental Pesticide Control Act, an amendment that greatly strengthened the original Federal Insecticide, Fungicide and Rodenticide Act.

One resource used in agriculture, namely, energy, received considerable attention in the 1970s. Escalating prices and awareness of limited energy sources caused concern among farmers and all users of diesel fuel, gasoline, and natural gas. The problems for farmers were further compounded by escalating prices of fertilizers, which are energy-intensive products. Farmers' problems were exacerbated further by temporary shortages of natural gas in the 1976–1977 winter and of diesel fuel in the 1979 summer. These temporary shortages occurred during periods of critical need, that is, during grain drying and field operations, respectively. Farmers feared that impending energy shortages would coincide with periods of peak demand.

Recognition of energy problems, environmental concerns, and threats to soil and water resources resulted in federal studies and the release of important federal publications in the early 1980s. These included what I have come to think of as landmarks of federal awareness. Documents of special interest are the *Global 2000 Report to the President* (Barney 1980), *Report and Recommendations on Organic Farming* (USDA 1980), and *Impacts of Technology on U.S. Cropland and Rangeland Productivity* (OTA 1982).

However, awareness and concern do not necessarily translate into change and solutions. Agriculture still faces a number of problems, which are examined briefly in this chapter. More extensive data and approaches to solving these problems, both present and future, are presented in the following chapters.

Energy

The reality of limited agricultural resources arrived in full force in the 1970 decade, as the energy crisis caught the world's attention. As expected, awareness of limited resources resulted in price increases; political and other socioeconomic factors were also involved in the price issue.

The effect upon the farmer is easily seen in Fig. 1.1. Prices for energy derived from diesel, gasoline, and liquified petroleum gas increased several fold in the 1970s. As shown in Fig. 1.2, these price increases were a direct result of the energy crisis, and not due to increased fuel demand.

All agricultural inputs derived from petroleum increased in price during this period. The effect upon fertilizer was especially dramatic (Fig. 1.3), although some of the price increase resulted from heavier usage of nutrients (Fig. 1.4). The cost of other farm inputs increased as a result of inflationary pressure fueled by increasing energy costs (Fig. 1.5).

Collectively, the energy effect drove up production expenses of the farmer but did not improve net farm income (Fig. 1.6). Indeed, in terms of constant dollars American farmers were worse off at the end of the 1970s than at the beginning (Fig. 1.7). The cost–price squeeze encountered by farmers in the late 1970s has continued into the 1980s, resulting in credit crunches and foreclosures for some financially troubled farmers.

While these energy-related costs are troublesome now, the future is especially worrisome. Even more disturbing is the fact that if a U.S. diet was fed to the current world population of 4 billion, and produced by U.S. farm

% of 1967 Prices

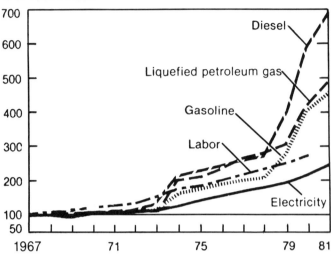

FIG. 1.1. The alarming increase in the costs of energy inputs during the 1970s had a significant impact on U.S. agriculture, since farming uses more petroleum than any other single industry.
From USDA (1982)

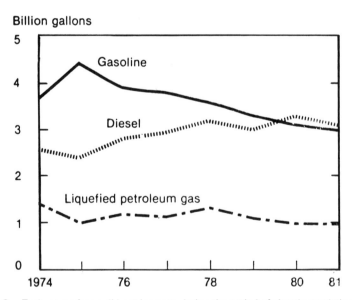

FIG. 1.2. Fuel use on farms did not increase during the period of sharply escalating energy costs. The reversed curves for gasoline and diesel fuel reflect a trend toward increased usage of farm machinery utilizing diesel fuel rather than gasoline.
From USDA (1982)

4

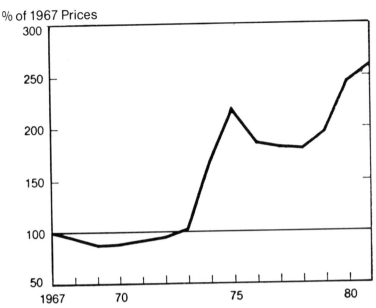

% of 1967 Prices

FIG 1.3. Fertilizer prices, highly dependent on energy costs, increased steeply in the last decade. The dip in price resulted from decreased fertilizer demand and is attributed to high prices in 1975 coupled with uncertain income prospects and wet field conditions in 1978.
From USDA (1982)

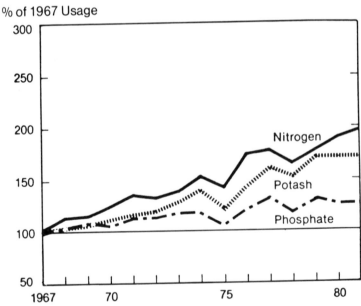

% of 1967 Usage

FIG 1.4. Nutrient usage increased modestly during the 1970s. The decreases in 1975 and 1978 were explained in Fig. 1.3. The decrease in potash and phosphate usage from 1979 on resulted from greater than expected costs of farm inputs not compensated for by crop prices (cost–price squeeze) and tight farm credit.
From USDA (1982)

5

% of 1977 Prices

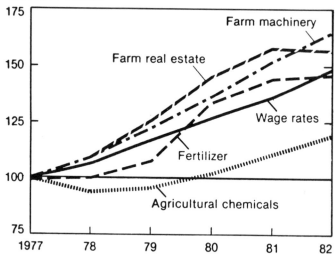

FIG. 1.5. Dramatic price increases in all farm inputs occurred during the last decade as a result of inflationary increases fueled by energy costs. During this same period the use of labor declined by about one-third, mechanical power and machinery use increased by one-third, and use of agricultural chemicals increased about two-thirds.
From USDA (1982)

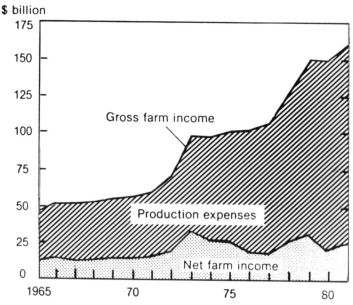

FIG. 1.6. Although gross farm income has increased about threefold since 1970, production expenses have also risen dramatically. Much of the increase in production expenses resulted directly from rising energy costs and indirectly from effects of these costs upon energy dependent farm inputs and inflation.
From USDA (1982)

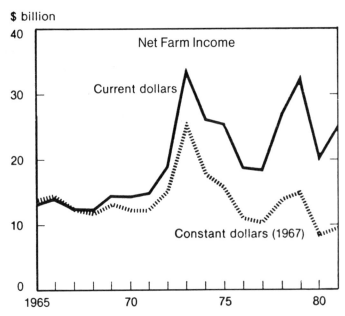

FIG. 1.7. Net farm income has not kept pace with inflation over the last several years. Farmers costs rose dramatically, at an average rate of 10% per year for 1977 through 1982, yet income increased only 7%. This cost–price squeeze resulted in large part from consumer resistance to higher prices, record crops, and slackened export demand.
From USDA (1982)

technology, the total known petroleum reserves would last only 13 years (Pimentel *et al.* 1975). This projection does not even consider population increases or nonagricultural uses of petroleum (heating, transportation, cooling), so the situation is even grimmer.

Water

Equally troublesome is another agricultural resource, water. Pressures on water resources stem from population increases, losses from pollution, and overuse of groundwater (CEQ 1983). Agriculture's water problems—unlike its energy problems—are more regional than national. For example, saltwater intrusion is occurring in Florida and Texas, salinity from recycled irrigation water in California, and groundwater depletion in the Texas high plains.

Other problems such as competition for water among farmers, home owners, and industry are in the beginning stages. This problem is surfacing in the Southwest and in the Northwest, where competition exists among agricultural, recreational, hydroelectric, and Indian interests. The Northeast is troubled with acid rain, the effect of which upon agriculture is not resolved. However, a growing body of evidence suggests a linkage between forest decline and acid rain (Klein and Klein 1985).

Agricultural activities may cause pollution of water. Runoff from agricultural sites has caused violations of water quality standards throughout the country (CEQ 1983). The pollution arises from sediment, animal wastes, agricultural chemicals, and irrigation practices. Agricultural use of water is the highest (83%) and also the most widespread cause of nonpoint source pollution.

Soil

Present agricultural practices in many cases lead to the "mining" of soil. For example, the production of 1 bu of corn in the Corn Belt consumes 2 bu of topsoil. Soil loss continues to be as serious today as during the Dust Bowl days of the 1930s (Comptroller General 1977; Pimentel *et al.* 1976).

Erosion problems are found throughout the United States. Especially serious problems exist in the Great Plains, where erosion from wind and water is substantial, not only on cropland, but also on pasture and rangeland. The soil losses from cropland alone are some 2 billion tons annually.

Losses of soil result in decreased productivity. Up to now these losses have been offset by increased fertilization. However, the increasing costs of fertilizers and long-term implications of their heavy usage decry the wisdom of this approach. Erosion also causes water pollution; sediment is rated as the biggest agricultural pollutant.

Transitional Agriculture

Awareness of the preceding problems by those involved in all aspects of agriculture has heightened rapidly and is beginning to stimulate changes in American agriculture. Essentially, agriculture is in an accelerated transition, with the long-range goal being a sustainable agriculture. Whether this goal— the elimination of agriculture's consumption and pollution of limited resources—can be reached is unknown. However, the attempts can only lead to a more sustainable agriculture, prolonging the future of American agriculture.

Before these various responses are examined briefly here and in detail in later chapters, one point needs to be made. Although the majority of agriculturalists (farmers, scientists, and policy makers) were not overly concerned with sustainability in the past, numerous individuals nevertheless were concerned and did make contributions. The difference now is that many more are involved, such that the shaping of a national trend is evident.

Some farmers were concerned about resource overuse and polluting activities. Their practices reflected those concerns and often departed from the traditional norm. If their departures were substantial enough, these farmers were sometimes labelled as "organic farmers." Still others used traditional practices, but managed them so as to conserve and not pollute resources; others mixed traditional and "organic" agricultural practices. Concerned agricultural scientists produced some new or modified procedures that resulted in conservation of resources or a reduction in pollution. Forward-looking policy makers shaped programs to increase soil and water conservation or to decrease pollution.

But the overall drive was for increased farm productivity as a national goal. Sustainability was a secondary and even an unplanned result. As pointed out earlier, national concern and action is at a crossover point, and the goal of sustainability is gaining considerable favor.

The response to the problem is highly variable. Interest in and an examination of "organic farming" and other alternative forms of agriculture is increasing. Changes are seen in traditional agriculture; variation here is considerable. Factors involved in the process of change include resistance to change, altruism, economic and regional contraints, profitability, skills, and knowledge. The choice is often complicated by lack of knowledge and/or poor dissemination of available information.

In some cases data is completely or partially lacking and research is needed. If data is not limiting, often its interpretation or dispersal is. Even when single practices are well characterized, the picture is incomplete, since the effect of an isolated component, once plugged into the agricultural system, is unknown.

Another problem concerns the need to improve communication. For example, the research of two separate scientists, if merged in an interdisciplinary project, might be far more productive than that of either alone. Another example is confusing terminology, especially in the area of alternative agriculture. Biological agriculture, conservation farming, ecological agriculture, environmentally sound agriculture, organic farming and agriculture, regenerative agriculture, and sustainable agriculture have all been used, sometimes without clear definition.

Clearly, the time has come for national action to decide what are the needs of sustainable agriculture, to provide funds, to direct and develop relevant programs, and to maintain communications with all those involved. With that in mind, we will now turn to the challenge that confronts us.

Meeting the Challenge

Choices and directions concerning the future of agriculture must be decided now if future generations are to be assured an adequate food supply. We have a moral obligation to future Americans; we must not fail them, no matter how costly and difficult the decisions become.

We need more research directed at regional and national agricultural sustainability, but not necessarily in the sense of isolated components. More interdisciplinary research is required, with a view toward understanding and predicting interactions of components and subsystems of a sustainable agricultural system. Only then can the true potential of a new practice or application be ascertained.

National and regional clearinghouses are needed to coordinate, receive, assess, and distribute all research results. Research should be at traditional institutions, such as the U.S. Department of Agriculture (USDA) and state agricultural experiment stations, and at newer alternative research centers. Examples of the latter include the well-known Rodale Research Farm in Pennsylvania and the New Alchemy Institute in Massachusetts.

Agricultural educators and extension personnel must be constantly updated on new developments, if information on sustainable agriculture is to reach and benefit current farmers and be available for training future agriculturalists. Information needs to reach those in need, the farmers, and be communicated in an understandable form. Demonstrations with actual farms will help greatly. This information must be targeted to specific regions to take into account the distinct regional differences in U.S. agriculture.

The public must be made to understand the seriousness of the failure to move toward a sustainable agriculture. Once the future specter of food shortages, food priced out of reach for many, and the possibility of strife and war brought on by impending starvation is realized, the public will support more agricultural funding. The present public image of a hugely successful American agriculture, widely preached, will be difficult to overcome, but it must be done.

Meeting the challenge of developing a sustainable agriculture sounds overwhelming, in terms of funding, manpower, and facilities. We do have the resources, however, to meet the challenge, to mount a Manhattan-type project for the future salvation of agriculture.

The manpower and facilities already exist within the USDA and the allied Land Grant universities. The latter could be the focus of regional research programs concentrating on sustainable agriculture, conducted by agricultural research scientists at the state agricultural experiment stations and the state colleges of agriculture. Cooperative research activities with private research centers dealing in alternative and new agricultural technologies should be encouraged. The extension service at the land grant universities should give high priority to communicating the results of sustainable agricultural research to farmers and the public. Demonstrations could be conducted on experimental farms, and training sessions held for farmers to assure information transfer. At the same time that agricultural faculty train future agriculturists, they could pass on the latest knowledge of sustainable agriculture. When needed, they could share in the training of established agriculturalists through participation in the workshops set up by the extension service.

While USDA research laboratories in various regions might participate in regional sustainable agriculture research, their primary objective would be on research applicable to the national needs for a sustainable agriculture. Such projects might include genetic engineering of plants to fix nitrogen directly from the air to lessen fertilizer needs, to engineer plants and animals with greater productivity and increased insect and disease resistance, or to adapt plants to marginal lands. Other projects might aim at energy self-sufficiency through the combined use of solar energy and biomass fuels. These are but a few of the many possibilities.

Within each agricultural region, one USDA laboratory could be designated as the coordinator of regional research directions and the clearinghouse for regional activities in sustainable agriculture. Research results could be gathered from the private, state, and federal activities in the region, examined and summarized, and coordinated information distributed to those involved in dissemination of regional information.

Regional information would also be fed to a national clearinghouse to glean data applicable to national needs. The national clearinghouse would coordinate those research activities of the various regional USDA laboratories that were designated as serving national need. The national clearinghouse would also conduct national sustainable agriculture research, but in a complimentary, not overlapping manner. Results from these projects would be analyzed and tailored for different regional applications and sent to the appropriate regionable USDA clearinghouses for distribution. The ideal choice for the national clearinghouse would be the USDA operation in Beltsville, Maryland.

The question of funding remains. Research, extension, and educational activities relating to sustainable agriculture should be assigned the highest priority and existing funding redistributed to reflect this priority. Additional funding could be sought from the federal government. Effective communication would be needed to assure that public and federal support would be forthcoming. In the final analysis, a sustainable agriculture research program can be related to national security in terms of increasing food exports and reducing the risk of internal strife and global warfare stemming from hunger and starvation.

The possible accomplishments of a national, concerted effort would be considerable, in view of what has been achieved with the present less-than-unified attempts. For example, definite improvements in energy conservation have occurred in crop and animal production, as well as food production. The most notable success, in terms of new technology and rapid adoption, has been in the greenhouse industry, where savings up to 50% are possible and actual energy conservation is realized by many in comparison with previous practices. These are specialized, often technical applications, which have reached the grower through well-organized communication routes. In other cases simple alterations of basic practices have trickled down to the farmer, such as changes in tractor operation or running of animal facilities. While far from highly technical, these changes have conserved energy. Both technical and simple changes have conserved energy during food processing.

Soil and water conservation are also becoming more widespread, but not nearly as much as energy conservation. While some effective practices are available, cost often limits their acceptance. Some notable exceptions are no-tillage systems and crop rotations involving corn. Both these practices slow erosion, save energy, and limit pollution.

One successful approach to limiting pesticide pollution is integrated pest management (IPM). Besides reducing pollution, it saves some energy and has been shown to be a viable, economical alternative to pesticides. Although more demanding in terms of knowledge and management, IPM is being increasingly adopted, especially since many insects are becoming resistant to pesticides presently used.

Several areas of active research also offer considerable promise for improving the sustainability of agriculture. As expected, they are highly technical and may be years in the future.

With a view towards a comprehensive discussion, all of these trends, successes, and past, present and future developments pertaining to sustainable agriculture are presented in this volume. A convenient division is into alternative and traditional farming. In Chapter 2 the most successful form of alternative agriculture, organic farming, is examined. Organic farming is well established, viable, and becoming well characterized. It also shows a definite advantage over traditional agriculture in terms of sustainability. Chapter 3 and the following chapters deal with past, present, and future agricultural modifications, some in common with organic farming and others not, that can help transform traditional agriculture into a more sustainable agriculture. Only time, money, and a national effort will make it truly a sustainable agriculture.

The discussion that follows varies considerably in terms of format, level, and appeal. Such an approach is unavoidable; it truly reflects the diverse nature of the available information. A review style is maintained wherever possible, with appropriate analyses of existing information, references to more detailed sources, and indications of needed research. Sometimes the material appears somewhat simple, almost how-to; this occurs when actual, simple practices are in the process of assimilation by farmers. Other times the material may be highly technical, for example, the discussions of energy conservation in greenhouses and computer applications. Again, this reflects the actual situation of contemporary adaptation.

This material will not interest everyone to the same degree. Some will appeal to those involved in crop or animal production, or perhaps greenhouse managers. In some cases it may interest agricultural researchers, or food processors, or extension workers. But that's the way it should be. The specific applications of sustainable agricultural principles will be relevant to a widely diverse group sharing one overall objective.

References

BARNEY, G. O. 1980. The Global 2000 Report to the President: Entering the Twenty-First Century. Vol. 1. GPO, Washington, DC.

CEQ. 1983. Environmental Quality 1982. Thirteenth Annual Report. Council on Environmental Quality, Washington, DC.

COMPTROLLER GENERAL. 1977. To Protect Tomorrow's Food Supply, Soil Conservation Needs Priority Attention. Rept. CED-77-30. U.S. General Accounting Office, Washington, DC.

KLEIN, D. T. and KLEIN, R. M. 1985. The growing case against acid rain. Garden 9(2), 22–27.

MEADOWS, D. H., MEADOWS, D. L., RANDERS, J. and BEHRENS III, W. W. 1972. The Limits to Growth. Universe Books, New York.

MEADOWS, D. H., MEADOWS, D. L., RANDERS, J. and BEHRENS III, W. W. 1974. The Limits to Growth. Rev. ed. New American Library, New York.

OTA. 1982. Impacts of Technology on U.S. Cropland and Rangeland Productivity. Office of Technology Assessment, U.S. Congress, Washington, DC.

PIMENTEL, D., DRITSCHILO, W., KRUMMEL, J. and KUTZMAN, J. 1975. Energy and land constraints in food-protein production. Science 190, 754–761.

PIMENTEL, D., TERHUNE, E. C., DYSON-HUDSON, R., ROCHEREAU, S., SAMIS, R., SMITH, E., DENMAN, D., REIFSCHNEIDER, D. and SHEPARD, M. 1976. Land degradation: Effects on food and energy resources. Science 194, 149–155.

USDA. 1980. Report and Recommendations on Organic Farming. U.S. Dept. Agric., Washington, DC.

USDA. 1982. 1982 Handbook of Agricultural Charts. Agric. Handbook 609. U.S. Dept. of Agric., Washington, DC.

USDA. 1983. Fact Book of U.S. Agriculture. U.S. Dept. of Agric., Washington, DC.

2

Organic Farming

If we were to play a word association game and someone said "sustainable agriculture," the reply would more than likely be "organic farming." This point raises some interesting questions. Are they associated practices? Are they equivalent? To answer this question, we must turn to working definitions.

A sustainable agriculture system is one in which the goal is permanence achieved through the utilization of renewable resources. The permanance sought is not static, but dynamic. The latter is necessary because some resources (e.g., population increases and cost increases for a diminishing supply of fuel energy) are not truly controllable. Hence, the need is for a dynamic permanence—one that permits the attainment of optimal production through the application of increasingly efficient techniques.

Basic elements of sustainable agriculture are the conservation of energy, soil, and water. Essential to the achievement of conservation are agricultural practices that are directed toward renewability and nonpollution of resources. A renewed resource that is contaminated with pesticides and unusable or a polluted environment that reduces crop productivity is counterproductive.

Such concepts do not imply a return to primitive or nonindustrialized agriculture with the complete abolition of agricultural chemicals. More efficient methods undoubtedly will depend on increased technology, a technology that is energy efficient and nonpolluting, or nearly so. Return to a chemical-free, less technological agriculture would be costly in terms of increased human labor and, most likely, would result in decreased agricultural productivity. Essentially, sustainable agriculturists will have to balance resource conservation and environmental maintenance against maintenance of optimal productivity and minimal labor input.

Agricultural chemicals and mechanization would still have a place in sustainable agriculture, but not as before. Use of less chemical fertilizers would reduce energy demands and pollution; fertilizers would be supplemented with legumes, crop residues, green manures, and readily available, local supplies of organic wastes. Their utilization would help further to reduce costs and energy demands, remove a source of potential pollution, and, more importantly, renew soil organic matter. Practices to control erosion would be maximized to make soil a renewable, rather than a minable resource.

Pesticides would still be used, but only as a component of integrated pest management systems that rely primarily upon effective natural controls. Pesticides would be used only when natural controls were not effective and selection would be for the least toxic of those available. An IPM program would maintain pest control, but reduce pollution of other agricultural resources and minimize detrimental effects on environmental quality. Beneficial organisms would be less likely to be destroyed and the induction of resistance in pests would be minimal.

Growth regulators and mechanization would be used in sustainable agriculture to reduce labor needs. However, in comparison with traditional practices, the tendency would be toward a conservative application of mechanization and search for more efficient machinery and practices to reduce energy consumption. The thrust would be to maintain or improve agricultural productivity while ever mindful of labor input and energy consumption.

Organic agriculture has much in common with sustainable agriculture. The same stress is placed upon the use of renewable resources, the need for conservation of resources (energy, soil, and water), and the maintenance of environmental quality. Organic farmers have no desire to eliminate mechanization. They are every bit as aware of productivity and economics as the conventional farmer.

The one major difference is in the use of agricultural chemicals. Sustainable agriculturists think in the sense of supplementation; organic agriculturists think in the sense of replacement. Fertilizers synthesized from chemicals are avoided. Nutrient requirements are met solely with animal and green manures, legumes, organic wastes, and mineral-bearing rock dusts. Pesticides are replaced by mechanical cultivation, mulches, biological controls, and other forms of natural control.

The problem of distinguishing between the two agricultural systems arises with borderline cases. Some organic farmers resort to the use of agricultural chemicals when organic practices fail or give unsatisfactory results, although their use is very selective and sparing. Studies by a USDA study team on organic farming concluded that organic farmers were spread over a wide spectrum (USDA 1980). The USDA (as well as borderline organic farmers) considered borderline cases to be practitioners of organic farming, even though they departed occasionally from purist practices.

From the preceding discussion it is clear that organic farmers are practitioners of sustainable agriculture, but not all sustainable agriculturists are organic farmers. Regardless of their approaches to sustainable agriculture, both are motivated by the following concerns:

- *Energy*—Increased costs, variable availability at critical times, heavy reliance on, future limitations
- *Agricultural Chemicals*—Increased costs, heavy reliance on, environmental damage, health hazards, increased resistance of insects and weeds to
- *Soil*—Losses of organic matter, decreased productivity and tilth, erosion
- *Water*—Pollution, future limitations
- *Future*—Sustainability of agricultural production and profits

The objectives and practices of sustainable agriculture owe much to organic agriculture. Concerns leading to the concept of sustainability originated with organic farmers and their advocates. In time these "organic" concerns spread into the conventional farming, public, and governmental domains. These concerns became the impetus for a serious reformation of agriculture during the 1970s, a new direction in which much of the organic agriculture philosophy became integrated with the high productivity practices of conventional agriculture. The new merger was deliberately devoid of the excesses of chemical agriculture, essentially a merger of moderation.

Since organic agriculture preceded sustainable agriculture and embodies much of its philosophy, it is productive to analyze the accumulated data of organic agriculture. Many of the practices and conclusions about the value of organic agriculture have been documented over a lengthy time and are applicable to sustainable agriculture. Analyses of organic agriculture include those of Beradi (1978), Eberle and Holland (1978), Lockeretz et al. (1976, 1978, 1981), Oelhaf (1978), Parr et al. (1983), Power and Bezdicek (1984), Roberts et al. (1979), USDA (1978, 1980), and Vail and Rozyne (1978). Emphasis in this chapter is placed on results derived from the combined practices inherent in organic farming. Some of these practices may already be shared by or be applicable to conventional farming. Analyses of individual practices, including actual and/or projected results in the context of the transition of traditional to sustainable agriculture, are presented in the following chapters.

Prevalence, Size, Common Practices

No reliable data exist for the actual number of U.S. farmers who practice some or all organic farming. In 1980, the USDA estimated there were at least 11,200 organic farmers if a strict definition was used, and at least 24,000 if farmers who practice organic techniques on part of their cropland are included. The USDA belief is that the actual number will be much larger when complete documentation becomes available (USDA 1980).

Contrary to popular belief, not all organic farms are small. They range in size up to 567 ha (1400 acres); combined conventional/organic farms of just conventional farms. The advantage of organic farms during drought un-doubtedly derives from the emphasis placed upon replacing or increasing soil organic matter. Organic matter contains micropores and is noted for increasing the water retention properties of soils.

Better research studies are needed, as stated previously, to assess the long-range productivity of organic farms under the variable conditions en-countered throughout the United States. Still it is reasonable to conclude from the available data that yields on organic farms during drought are comparable to or slightly better than those on conventional farms. As such, organic farming practices have potential value in areas of existing dryland farming, or in areas where irrigation is becoming limited by cost or water problems.

Under ideal conditions organic farming yields would vary from moderately less than to nearly comparable with conventional farming yields. Crops heavily dependent upon fertilizers and pesticides—in particular, corn and wheat—would yield somewhat lower on organic farms. The difference with less demanding crops, such as soybeans, would be smaller, and essentially un-noticeable with crops of minimal demand, such as oats and hay. Conventional farmers switching to organic practices would experience larger yield dif-ferences than those reported until soil organic matter and nutrient levels became well established through cumulative organic techniques. Recently established organic farmers confirm this conclusion (USDA 1980).

One caution is needed here. Conclusions about the desirability of organic farming should not be based strictly upon yield. The energy savings dis-cussed previously would more than compensate monetarily for the observed yield decreases. Other factors must be considered, such as current social or political views about organic agriculture in the context of resource sus-tainability and pollution. The effect of these views is difficult to quantitate, but they are likely to become a stronger incentive in the future. Such a force could lead to more organic farms and/or an increased acceptance of some organic methods on conventional farms. An interesting possibility for con-ventional farmers might be the adoption of organic practices emphasizing the maintenance of soil organic matter coupled to decreased use of agri-cultural chemicals. Such a modified farm should experience no reductions in yield, have minimal problems in terms of pollution, and be viewed favorably in the social and political sense.

Economics

Limited data is available on the economic performance of organic farms. The present consensus is that net returns from crop production can be similar for some organic and conventional crop–livestock farms (USDA

---

over 2429 ha (6000 acres) exist. The success of these farmers indicates that size is not a stumbling block (this will be covered later under economics). The majority of organic farms are owner operated; they are located in all regions except the South, where certain factors (climate, soil, insects, market conditions) limit organic agriculture. Organic farmers represent all ages and levels of education and experience (USDA 1980).

Certain cropping and cultural practices are an integral part of manage-ment on an organic farm. Common elements include legume-based rota-tions, green manures, crop residues, cover crops, animal manures, rock dusts, and small amounts of organic fertilizers. Common crops are forages, vegetables, small fruits, and small grains. Tillage is primarily with chisel plow or disk and often is shallow (3–4 in.). Insects are controlled through selective rotations, biological methods, and organic insecticides. Weed control is usu-ally through proper timing of tillage and planting, crop rotations, and mow-ing. Mechanization is not shunned and efforts toward soil and water conser-vation are excellent (USDA 1980).

Energy Usage

An examination of the energy usage by organic farmers indicates that their operations are more energy efficient than those of conventional farmers. As expected considerable savings are possible by the elimination of agricultural chemicals, especially fertilizers, which are highly energy intensive. This saving of energy is partially offset by increased energy consumption for additional mechanical cultivation and somewhat reduced yields (see next section, Crop Productivity). Documentation, though not extensive, does exist in the liter-ature.

Lockeretz *et al.* (1976, 1978, 1981) conducted intensive research on corn production in the Midwest. Their study spanned 5 years and compared con-ventional vs. organic farming on several parameters. While the farms com-pared were matched as closely as possible in terms of soil, size, and types of crops, they did differ somewhat in the crop mix. The conventional mixed grain–livestock farms had the following crops in descending order of acreage: corn, soybeans, hay, oats, and wheat. The hay and soybeans were reversed for the mixed grain–livestock organic farms because of their rota-tional requirements. About 10–15% more corn or soybeans was grown on the conventional farms, but about 10% less oats and wheat, than on the organic farms. Because of this relative difference in crop mix, the results were aggregated for all the cropland on each farm and the energy value was expressed per unit value of production. The results indicated that the conven-tional farms required 2½ times as much energy per unit value of production than did the organic farms. Most of the difference was attributed to corn.

Berardi (1978) looked at the energy profile for the production of wheat on organic and conventional farms in New York and Pennsylvania. The conventional growers required about 43% more energy; however, once an adjustment was made for yield differences, the difference in energy use was only 18%. Further studies with wheat indicated net energy savings of 15–47% were achieved by organic farmers, and a 25% energy savings was noted for barley (USDA 1980).

Crop Productivity

Several problems exist in assessing yield comparisons between organic and conventional farms. First, there is a paucity of yield data from long-term, well-designed experiments. Replicated, paired experimental plots (organic vs. conventional farming) must be examined over long periods in various geographical locations and under many different conditions and practices before valid comparisons can be made. In fact, such experiments are extremely difficult for several reasons:

- Large numbers of variables—many different soils; variations in climate; variations in crops; sources of organic matter/fertilizer; variations with insects and diseases
- Large range of organic and conventional farming practices—tillage variations; plant protection variations; timing differences for planting, fertilizing, and control of pests; rotation practices
- Prolonged length of time and large number of experiments needed to assess system stability and statistical significance of data

The data that do exist are subject to criticism. Some experiments use replicated, small-plot comparisons of individual treatments. These studies can be challenged fairly as an inadequate representation of organic farming as actually practiced, in that they do not reflect the reality of integrated practices with their additive and synergistic effects.

Other experiments are short-term comparisons with large plots, actual fields, or entire farms, but with small numbers of comparisons. Possible criticisms are that substantial errors of undeterminable nature are quite possible due to the difficulty of controlling or matching variables such as soil or climate, and the difficulty of accurate measurement of yields in large areas. Certainly they do not reflect yields indicative of the many possible variations on the national level.

With these limitations in mind, data on yield comparisons can provide some insight. However, these yield data are not necessarily a reliable indicator of the long-term performance of organic agriculture, nor are they applicable to any organic farm, regardless of variations in crop, soil, and climate conditions.

The most complete comparative study to date of crop yields on organ and conventional farms was by Lockeretz et al. (1981), which was discus in the section on energy conservation. The yield data was based both up farmers' reports and actual field measurements. These data were deri from neighboring organic and conventional farms of the same soil type, same crop variety, and the same or close planting time. Yields on orga farms were 8% lower for corn, 5% lower for soybeans, and 43% lower wheat than on conventional farms. Berardi (1978) also noted a 22% red tion in wheat yields on organic farms.

Sufficient data on corn were available for Lockeretz et al. (1978) to c duct a more detailed examination of yield. When growing conditions v better than average, corn yield differences were generally higher than average 8%. When growing conditions were poorer (e.g., during droug corn yields on organic farms approached and even exceeded those on c ventional farms. Yields of corn on three of the 26 organic farms were hig than 10 MT/ha (159 bu/acre). These are good yields for even a conventio farm.

Data on average yields of important crops on organic and conventi farms in the Midwest are shown in Table 2.1. The years represented v during a period of serious drought in the Midwest which affected yields in Corn Belt. As observed by Lockeretz et al. (1981), the combined data ap to suggest that during periods of drought corn yields on organic farms t to approach those on conventional farms and that yields of soybeans, wh and oats on organic farms can be reasonably competitive with those

Table 2.1. *Average Crop Yields on Midwestern Organic and Conventional Farms, 1973–1976*

Crop	Bu/acre[a,b]	
	Organic	Conventional
Corn	77.9 ± 5.4	80.6 ± 7.6
Hay	4.8	3.7
Oats	58.3 ± 3.3	57.0 ± 4.7
Soybeans	30.0 ± 2.9	29.9 ± 4.0
Wheat	31.4 ± 3.8	34.4 ± 4.1

Source: Klepper et al. (1977), Lockeretz et al. (1978), and Roberts et al. (1979).

[a]Each average is based on 6 or 7 yield reports derived from both annual averages of actual comparative experimental farms and county data. Hay based upon limited data.

[b]Metric equivalents: 16 bu/acre = 14 hectoliters/ha, or about 1 MT/ha.

1980). More data will be required before this conclusion can be accepted as generally valid regardless of farm size, type, and all crops. Several individual studies are discussed in this section.

Eberle and Holland (1978) compared the crop returns on three pairs of conventional and organic farms in Washington. The net return per acre was 33% higher on the conventional farms. The data have been questioned because of the small sample size and the difficulty of pairing factors. In another study in the Northwest area (Kraten 1979), six organic farms, some of which used small amounts of chemical fertilizers, showed a 22% higher net return than conventional farms with similar crops. In the western Corn Belt, Roberts et al. (1979) compared 15 organic crop–livestock farms against conventional farms. The majority of the organic farms had net returns that surpassed those of the conventional farms.

Lockeretz et al. (1981) conducted two studies on the economic performance of organic farms. The first study, from 1974 to 1976, covered 14 organic crop–livestock farms with an average size of 172 ha (425 acres) in five states: Illinois, Iowa, Minnesota, Missouri, and Nebraska. The organic farms were compared with nearby conventional crop–livestock farms of similar size and soil type. The second study included 23 organic farms in 1977; 19 of these were repeated in 1978. In this study the organic farms were compared with a conventional farm profile determined from data derived from county and federal sources. The organic farms in the second study were located in Illinois, Iowa, and Minnesota and averaged 95 ha (235 acres); their livestock was restricted to beef and/or hogs and their soil had been mapped. Conventional farms of a similar nature in the same localities had a similar average size (96 ha).

The data collected included rates of seeding, choice or cultivars, fertilizer applications, manure and other applications, field operations (tillage, cultivation, harvesting), and yields. The results were expressed as cropland averages and covered the following: cultivated crops, crops used in meadow rotations and for soil improvement, and crops that failed. Market values of the crops were determined based upon reported yields and the state market price, regardless of the ultimate use of the crop. Some crops were sold and others used for livestock. The premium prices offered for some organic crops were not reflected in the calculated values. Fixed or overhead operating expenses were assumed to be the same for the organic and conventional farms, based upon their similarity in size and machinery. Therefore, only direct or variable operating expenses were included: seeds, materials, fuel, labor, repair of equipment, and drying of crops. Labor charges were calculated at prevailing rates, even when work was performed by the farm family.

From 1974 to 1977 the average net returns of organic and conventional farms were within 4% of each other. Organic farms had a lower market value by 6–13%, but their operating costs were less by a similar amount. However, in 1978 a change occurred. Although the operating costs for the organic farms continued to be lower and the net return per hectare was comparable with that in 1974 to 1977, their income per hectare was 13% below that of the conventional farms. The reason for this was that the average net return per hectare for the conventional farms was 21% higher in 1978 than in 1974–1977. Thus, the organic farms did not perform any worse in 1978 than in the previous study years; rather, the conventional farms performed significantly better.

The only discernible difference in these study years was that the severe drought conditions, which existed in 1974–1977, were ameliorated considerably in 1978. During the drought years (1974–1977) the yield benefits of the agricultural chemicals used on the conventional farms were minimal because other factors (primarily soil moisture) limited yields. However, in 1978 when growing conditions were more favorable, conventional farms appeared to have a competitive advantage over organic farms, as noted previously. Additional studies are needed to resolve this point. Nonetheless, sustainable agriculture, which employs both organic and conservative conventional practices, should not be at a disadvantage during periods of favorable growing conditions.

The USDA (1980) formulated Midwest farm budgets based upon available data in order to compare economically crop rotations on organic farms and continuous conventional crop practices. The analysis assumed that yields on organic farms were 10% lower. Rotations tie up part of the cropland with forage legumes, such as alfalfa; on conventional farms this land would be producing either corn or soybeans. Since corn and soybeans command a higher price, potential income is reduced in proportion to the amount of land tied up in forage legumes.

Three rotations were compared with conventional continuous corn and soybeans in the USDA study. The conventional system produced 21.6% more income than the 7-year rotation (alfalfa, alfalfa, corn, soybeans, corn, soybeans, and oats), 24.6% more income that the 4-year rotation (alfalfa, corn, soybeans, oats), and 30.4% more income than the 5-year rotation (alfalfa, alfalfa, corn, soybeans, oats). In assessing these results, some factors must be considered. For example, in the 7-year rotation corn and soybeans are produced on only 57% of the total acreage annually. Yet the organic farm yielded an income not 57% but 82% that of the conventional farm with continuous corn and soybeans. Energy savings on the organic farm helped reduce the income difference between the two systems. In the long run the profit difference could decrease and even reverse if the price of energy increases substantially and the yields on conventional farms decrease as soil resources dwindle. The organic farmer is essentially turning part of his potential income into renewal of the soil (by adding organic matter) in order to assure sustainability of future crop production. The conventional system maximizes present income and is not as concerned about viewing soil as a long-term investment. In addition, rotations could become increasingly cost effective if demand for forage crops becomes stronger.

Projections about the economic impact of shifting from conventional to organic farming have been made by the USDA (1980) and Olson and Hardy (1979). Some interesting conclusions were reached. Small farms at the time of the studies accounted for 69% of the total farms and had annual sales of less than $20,000. These farms brought in 11% of total farm cash receipts. All these small farms could shift to organic farming with little economic impact upon the U.S. economy; however, a total shift to organic farming would have a major impact on the U.S. economy.

If agriculture shifted entirely to organic practices, crop production would still meet U.S. needs, but surplus for exports would be essentially eliminated. Food prices would rise significantly. The cost of the shift would be spread over 3–5 years, the time required for a shift from organic to conventional farming. The effect of a more gradual increase of organic farming in the future is difficult to assess based upon the present inadequate data base. Major savings in energy through decreased use of agricultural chemicals would only be possible if the large conventional mixed crop–livestock farms and the large farms producing the major crops switched to organic practices.

Labor

A number of variables determine the intensity of labor required in a farm operation. These include the types of crops and livestock, soil topography and type, the size and type of equipment and machinery, type of agricultural practice, and the efficiency of the management and labor. As might be expected, labor intensity varies from organic farm to organic farm and from conventional farm to conventional farm. This variation makes it somewhat difficult to make a general assessment of the comparative labor differences between organic and conventional farms.

Labor usage on organic and conventional farms is shown in Table 2.2. Although information is limited, it appears that organic farms require more labor for management than conventional farms. The increased labor can be attributed to the control of insects, diseases and weeds through mechanical and natural methods as opposed to chemical pesticides. The difference can be kept to a minimum, if hand weeding or hand picking of insects is not used. Such practices were responsible for a 300% difference in labor requirements of two organic grain farms in Washington (Table 2.2). The use of horses instead of powered vehicles also increases labor requirements (Table 2.2). On comparable organic and conventional farms with similar mecha-

Table 2.2. *Labor Requirements on Organic and Conventional Farms*

Crop	Location	Hr/ha(hr/acre)	
		Organic	Conventional
Corn, soybeans, small grains	Corn Belt	8.2(3.3)	7.9(3.2)
Corn, soybeans, small grains	Corn Belt	7.4(3.0)	6.9(2.8)
Grain	Washington	14.0(5.7)[a]	1.5(0.6)
		4.7(1.9)	3.2(1.3)
Wheat	New York,	21.0(8.5)	8.9(3.6)
	Pennsylvania	13.1(5.3)[b]	

Source: Lockeretz *et al.* (1976), Klepper *et al.* (1977), Eberle and Holland (1978), and Berardi (1978).
[a]Insects controlled by hand labor.
[b]Same group but excluding organic farmers using horses in place of mechanization.

nization and production, the labor input on organic farms, according to Lockeretz *et al.* (1981), was 12% more per unit value of crop production, or 3% more per unit value of land. They concluded that differences in labor reflected the basic difference in cultivation and crop mixes (less corn and soybeans, more forage legumes) practiced by organic farmers.

Additional labor information in different forms is also available. Klepper *et al.* (1977) showed that Corn Belt organic farms required 19.8 hr/$1000 of corn, soybeans and small grains produced as opposed to 17.8 hr for conventional farms. Lockeretz *et al.* (1976) in a similar study for the same crops gave a figure of 0.06 hr/bu and 0.05 hr/bu for organic and conventional farms, respectively. Oelhaf (1978) showed that most vegetables and fruits required more production labor when produced organically than when produced conventionally. The only exception observed by Oelhaf (1978) was intensive tomato production for the fresh market, which required similar labor inputs for organic and conventional systems. Another exception was observed by Roberts *et al.* (1979), who examined labor costs of organic and conventional western Corn Belt farms. They concluded that labor costs for corn, oats, and wheat were less in most cases on organic farms, but more for soybeans.

The need for additional labor on most organic farms does pose some problems. Good farm labor cannot always be found readily, as indicated by a study of organic farms in Maine by Vail and Rozyne (1978). The USDA (1980) feels that labor is a major limitation to increasing the size of some organic farms, especially for vegetable crops requiring hand weeding, and for increasing the number of farms shifting from conventional to organic farming.

Conservation and Environmental Quality

The idea of conservation and maintenance of environmental quality is inherent in the agricultural practices utilized by organic agriculturists. In practice organic farmers rely heavily upon crop and soil management practices that aid water infiltration, resist soil erosion, improve soil tilth and productivity, recycle organic wastes, and reduce pollution of the soil and water.

Soil

Conventional agriculture is exacting a heavy price in order to achieve its high rates of crop productivity. The price is "mining of the soil," or soil erosion. The system can only work so long under these conditions. Present estimates are that soil erosion is leading to a drop in crop productivity equivalent to an annual loss of 506,072 ha (1.25 million acres).

The organic farmer uses a number of practices that are extremely effective in reducing soil erosion. Continuous cropping is replaced by rotations that include meadow, legumes, and small grains along with row crops. This can result in a reduction of soil erosion. Cover crops, green manure crops, and reduced tillage help to conserve surface crop residues and to increase water infiltration. Cover crops also help prevent soil losses during the unproductive part of the growing season. Maintenance or increases of soil organic matter by application of manures and other organic wastes increases water infiltration and storage. These practices, therefore, decrease surface water runoff and wind blowoff, thus reducing losses of soil through erosion. Increased water storage and less runoff also reduce the contamination of water supplies with nutrients and pesticides (if they are used at all).

The effectiveness of these practices can be evaluated mathematically through the use of the C factor (cover and management factor) in the Universal Soil Loss Equation (Wischmeier and Smith 1978). Values for the C factor range from 0.001 for well-managed woodlands to 1.0 for continuous fallow; these values are directly proportional to differences in soil loss caused by differences in the management of soil and crops.

USDA (1980) analyses of organic farming practices have demonstrated their value for erosion control. For example, where rotations utilizing sod crops (grass and legumes) constitute 25–40% of the crops, soil erosion is 33–12.5% of that observed with continuous row crops and conventional tillage. Cover crops, when they follow harvested crops such as silage corn, potatoes, and most vegetables, are important in protecting the soil, since the preceding harvested crops leave little residue. Cover crops in such systems can reduce soil losses through erosion by 50%. The use of the chisel plow and disk by organic farmers, instead of the moldboard plow, can reduce soil erosion by 20–75%. The effectiveness of this practice results from the placement of crop residue at or near the surface.

Increased organic matter in the soil can also reduce the loss of soil. On the average, an increase of 1% in soil organic matter can decrease the potential for soil erosion by 10%. Such increases in organic matter are possible with continuous organic practices of manure applications and grass–legume rotations (Cooke 1977). Other soil conservation practices used by organic farmers include contour planting and plowing, terraces, and grassed waterways. The quantitative value of these practices was not evaluated in the USDA study of 1980.

The collective value of these and other practices has not been researched to any significant degree in reliable, comparative studies of organic and conventional farms. Lockeretz *et al.* (1981) estimated that organic crop–livestock farms in the Corn Belt had a loss of soil by water erosion about one-third less than that of conventional paired farms. This estimate was based on rotations and did not include the reduced tillage practices of the organic farmer, which would have increased the difference even more.

Besides preventing soil erosion, the preceding practices of organic farmers are the very management practices that have been advocated by the USDA and Land Grant universities for improving soil tilth and productivity (USDA 1980). Over a long period of time this type of agricultural management can increase aeration, aggregation, permeability, and water storage capacity; decrease compaction and crusting; and improve levels of beneficial soil organisms (insects, earthworms, and microorganisms).

Water

A number of organic practices increase water infiltration, thus increasing water availability over the short term (USDA 1980). This improvement arises from increases in surface crop residues provided after plowing out or under cover crops, green manures, and sod crops, and from reduced tillage practices. The maintenance and increase of soil organic matter also aids water infiltration. The effectiveness of these practices was evident in the findings of Lockeretz *et al.* (1981) that the drought in the Corn Belt during the mid-1970s caused yield reductions for conventional farmers but not for organic farmers.

Increased water-holding capacity can result from long-term increases in or maintenance of a high level of soil organic matter. Reliable assessment of the results from short-term changes in organic matter is difficult (USDA 1980). Soils most likely to show increased water retention from the addition of organic matter are sandy soils. Further studies are needed on the long-term advantages of organic farming practices in terms of soil organic matter and its effect on crop productivity.

Environment

Organic farming practices that reduce soil erosion also reduce surface water runoff and thus are less likely than conventional practices to result in contamination of water supplies with nutrients and pesticides. The organic practice of not using agricultural chemicals also reduces the environmental impact of farm operations.

Excessive use of chemical fertilizers can lead to pollution of groundwater with nitrogen and phosphorus, especially if associated with high rainfall or irrigation, shallow-rooted crops, and sandy soils. Organic farmers use little, if any, chemical fertilizers. Instead they rely upon recycling of nutrients in their management approach; these nutrients generally are not susceptible to leaching, even when applied excessively. Indeed, many organic farmers tend to operate at nutrient levels below the maximum needed for optimal yields (USDA 1980). Less nutrients and less leaching coupled with practices that reduce soil erosion by water runoff minimize nutrient flow from the farm to groundwater.

The effectiveness of organic practices in reducing nutrient flow from farmland can be attributed to the slow release of nitrogen from organic materials such as manures, compost, sewage sludge, and residues from cover crops, green manures, and rotations. The continual crop cover provided by rotations, cover crops, and green manures utilizes the released nitrate, leaving little for leaching away. In addition, the rotation of crops that require high amounts of nitrogen (e.g., corn) with crops that require little nitrogen (e.g., soybeans and alfalfa) helps to lower the average amount of nitrate available for leaching. Alfalfa has deep roots, enabling it to retrieve nitrogen past the root zone of other crops. Winter cover crops extract nitrate and soil water, thus reducing the usual loss of nitrates in winter when the land normally lies fallow.

Several studies have demonstrated that sod and cover crops do, indeed, reduce the chance of pollution from nutrient runoff. For example, losses of nitrates and phosphates were 300–600% less with a corn–wheat–clover rotation than with continuous corn (Stewart et al. 1975). The planting of corn in a ryegrass cover also resulted in a 50% decrease in runoff, as well as a 40-fold reduction in the loss of soil (as sediment) and of nitrate and phosphate (Stewart et al. 1975).

This discussion should not be construed to mean that no threat of pollution exists from organic techniques. Improper management of manures and sludges can cause problems. Incorrect storage of manures and sludge can lead to pollution and nutrient waste through leaching action. Improper application or use of contaminated sludges can lead to pollution of soil and water, and possible release of toxic compounds that can end up in the food chain. Excessive use of organic wastes can contribute to pollution of groundwater both during the growing season and after harvest. The excess nitrate not utilized during growth can be leached, and the slow release of nitrate after harvest continues to cause further problems. Runoff removal of some nutrients is possible if manures and sludges are applied to frozen or snow-covered fields.

In addition, adequate phosphorus (P) and potassium (K) must be applied to balance the nitrogen (N) added from either organic or chemical sources. In one study, much nitrate remained in the soil after harvest of wheat in a wheat–corn rotation fertilized with 120 kg N/ha (107 lb N/acre) when no P and K were added. In contrast, little N was found in the soil at depths up to 2 m when P and K were added at rates near 25 kg/ha (22.2 lb/acre). This problem should concern organic farmers, since the USDA (1980) study found that few of the surveyed Midwest organic farmers attempted to balance the nutrient budget with phosphorus or potassium. Lockeretz et al. (1976) also noted an average net deficit of 13.4 and 45.9 kg/ha (12 and 41 lb/acre) of P_2O_5 and K_2O, respectively, on organic farms in the Midwest.

Even larger accumulations of soil nitrate are quite possible with organic farming. For example, if a legume crop followed a wheat harvest, as would be likely in a typical rotation, the concentration of nitrate would rise because of nitrate fixation by the legume. The potential for nitrate contamination of groundwater would exist. Under these conditions a nonleguminous crop would be better, since oats, timothy, and rye can reduce leaching losses 40–60% (Stewart et al. 1976).

Pesticide pollution by organic farmers is minimal because they use non-chemical methods of pest control most of the time. Even if occasional lapses occur, their effective practices to reduce soil erosion and, hence, runoff would minimize the problem of pesticide contamination of groundwater. Further, their choice of pesticide would likely be the least toxic and most biodegradable.

Future Prospects

The available research data and the continued existence of organic argiculture in the face of economic reality suggest that at least this particular form of sustainable agriculture is viable. It is unlikely, however, that organic farming in its pure form should be or in the future will be the predominant form of sustainable agriculture in the United States. There are several reasons for this viewpoint.

Arguing against the widespread adoption of organic farming are the uncertainty of labor supplies, the probable loss of agricultural exports, and the reluctance of many conventional farmers to give up agricultural chemicals. On the other hand, the organic approach has much to offer the conventional farmer who wishes to practice sustainable agriculture. First, a definite reduction in energy costs accompanies use of organic practices. Conservation of agricultural resources, especially soil, also is assured. Maintenance of environmental quality can be achieved, even with some use of agricultural chemicals. Finally, reasonable productivity can be attained with organic practices.

The ideal solution would be an integration of the best from both organic and conventional farming practices, a new direction designed to sustain agriculture. Essentially, organic practices would be retained, but supplemented with judicious use of agricultural chemicals. Integrated pest management approaches would be used to control pests. Soil maintenance would involve blends of organic and conventional techniques designed to maintain or increase soil organic matter and fertility. The thrust would be to minimize energy usage and pollution, but to maximize conservation of resources. The extent and proportions of combined organic and conventional farming practices would be determined by personal belief, the desired levels of economic return, and the desired investment in agriculture's future.

Research Needs

Although much is known about organic farming, extensive research is needed to assess the full potential of this farming system as a form of or source of practices for sustainable agriculture. Research needs have been thoroughly established by the USDA (1980) and prioritized by Parr *et al.* (1983).

One area of immediate urgency is the investigation of organic farming not as a collection of individual practices, but as a collective practice, that is, a holistic research approach. The practices of organic farmers need to be examined for their interrelationships. For example, what are the interrelationships between organic waste recycling, energy conservation, nutrient levels and availability, and energy conservation? What are the effects of organic wastes coupled with rotations, cover crops, and the lack of agricultural chemicals upon the microbiology and biochemistry of the soil? How does organic agriculture compare with conventional agriculture in terms of economic return and several resource parameters, such as soil and water sustainability, environmental quality, energy conservation, and crop productivity?

Much field research over long periods of time will be required. Research must be done through an interdisciplinary approach, if the holistic examination is to succeed. Various experts will be needed in the areas of agronomy, biochemistry, climatology, ecology, economics, entomology, microbiology, pathology, physiology, and soil science.

In addition to the gathering of data, there will be a need for interpretation of the results and an assessment of what effect changes in practices have upon organic farming as a whole, but especially upon sustainability and environmental quality. Here the use of computer-assisted mathematical modeling will be mandated.

This will require the application of systems science to organic agriculture, that is, the resolving of organic farming into various components and subsystems. Each component will become a mathematical model, able to be described individually and in relation to the other models. The collective interaction of the models becomes the system, or a working simulation of organic farming. With this approach the models can be manipulated to predict the system's behavior when given different inputs. Systems science has been applied successfully in other areas of agriculture, such as integrated pest management (see Chapter 9). However, much work remains, if system science is to be applied to organic agriculture.

Many other areas require further study. For example, decreased yields are observed for 3–4 years during the conversion from conventional to organic farming. The causes, beyond weed problems, need to be identified in order to improve yields and economics during the transitional period. This aspect would appear to be especially important if conventional farmers are to be encouraged to become organic farmers, or to adopt specific organic practices.

A number of questions remain unanswered in terms of organic wastes. The inventory of organic wastes is presently incomplete, especially for industrial wastes produced by food processors, logging and wood product manufacturers, pharmaceutical producers, leather processors, and so forth. Data on all wastes are needed in terms of kinds, amounts, availability, and nutrient content.

Improved methods for handling each organic waste need to be developed. Is processing before land application necessary to reduce phytotoxicity or health hazards, or to enhance nutrient availability? Do any environmental or other constraints apply to long-term use of organic wastes? Can agricultural uses be improved in terms of labor, nutrient recovery, and economics? What are the effects of organic wastes upon soil properties, such as maintenance of organic matter, erosion control, tilth, water infiltration and retention, levels and activities of beneficial soil organisms, and fertility? Effects upon fertility are especially important; what is the relationship between application rates and frequency and nutrient availability? Soil tests will need to be improved considerably if these questions are to be answered.

Can the benefits resulting from application of organic wastes to soils be increased further? For example, there appears to be a synergistic effect when organic wastes are combined with chemical fertilizers. What are the combinations that produce optimal effects upon crop yields?

The relationship between energy conservation and use of organic wastes needs to be analyzed. The application of organic wastes to soils offers some energy savings, since fertilizer needs are reduced, and produces benefits resulting from maintenance of organic matter, such as erosion control. If some of these wastes were diverted to direct energy production, would the gain in energy offset the loss of benefits derived from use of wastes in the soil? What percentage can be diverted without harm to long-range soil productivity? What must be done to maintain soil productivity on biomass plantations?

Other areas need investigation. Can biological nitrogen fixation by legumes be improved, perhaps through selective breeding? Can it be enhanced by any changes in cultural practices? Can better nonchemical ways of controlling insects, diseases, and weeds be developed? Can plant breeders produce crops that are better adapted to organic farming?

Of course, results from all these avenues of research must reach the mainstream of agricultural activity, including education, extension, farmers, and the public. This is especially important for successful conversion or partial adaptation. Dissemination of research results must be improved substantially; organic farming practices have been poorly communicated and understood in the past.

One final thought. The amount of research suggested here is extensive and perhaps easily dismissed because it seems to only apply to organic farming. Although such research would certainly improve organic farming practices and perhaps increase the number of successful conversions, its most important practical outcome would be development of practices tailored specifically to improve the sustainability of conventional agriculture.

References

BERARDI, G. M. 1978. Organic and conventional wheat production: Examination of energy and economics. Agro-Ecosystems 4, 367–376.

COOKE, G. W. 1977. The roles of organic manures and organic matter in managing soils for higher crop yields: A review of the experimental evidence. In Proc. Intern. Seminar on Soil Environment and Fertility Management in Intensive Agriculture, Tokyo, Japan. Ministry of Agriculture, London.

EBERLE, P. and HOLLAND, D. 1978. Comparing organic and conventional grain farms in Washington. Tilth (Spring), 34–35.

KLEPPER, R., LOCKERETZ, W., COMMONER, B., GERTLER, M., FAST, S., O'LEARY, D. and BLOBAUM, R. 1977. Economic performance and energy intensiveness on organic and conventional farms in the Corn Belt: A preliminary comparison. Amer. Jr. Agric. Econ. *59*, 1–12.

KRATEN, S. L.   1979.   A preliminary examination of the economic performance and energy intensiveness of organic and conventional small grain farms in the Northwest. M.S. Thesis. Washington State University, Pullman.

LOCKERETZ, W., KLEPPER, R., COMMONER, B., GERTLER, M., FAST, S. and O'LEARY, D.   1976.   Organic and Conventional Crop Production in the Corn Belt: a comparison of Economic Performance and Energy Use for Selected Farms. Rept. CBNS-AE-7. Center for the Biology of Natural Systems, Washington Univ., St. Louis, MO.

LOCKERETZ, W., SHEARER, G., KLEPPER, R. and SWEENEY, S.   1978.   Field crop production on organic farms in the Midwest. J. Soil and Water Cons. *33*, 130–134.

LOCKERETZ, W., SHEARER, G. and KOHL, D. H.   1981.   Organic farming in the Corn Belt. Science *211*, 540–547.

OELHAF, R. C.   1978.   Organic Agriculture: Economic and Ecological Comparisons with Conventional Methods. Halsted Press, New York.

OLSON, K. D. and HARDY, E. O.   1979.   A National Model of Agricultural Production, Land Use, Export Potential, and Income Under Conventional and Organic Farming Alternatives. Center for Agricultural and Rural Development, Iowa State Univ., Ames. (Unpublished.)

PARR, J. F., PAPENDICK, R. I. and YOUNGBERG, I. G.   1983,   Organic farming in the United States: Principles and perspectives. Agro-Ecosystems *8*, 183–201.

POWER, J. and BEZDICEK, D. (Editors).   1984   Organic Farming: Current Technology and Its Role in a Sustainable Agriculture. Amer. Soc. Agron., Madison, WI.

ROBERTS, K. J., WARNKEN, P. F. and SCHNEEBERGER, K. C.   1979.   The economics of organic crop production in the western Corn Belt. Agric. Econ. Paper *1979–6*. Dept. Agric. Econ., Univ. of Missouri, Columbia.

STEWART, B. A., WOOLHISER, D. A., WISCHMEIR, W. H., CARO, J. H. and FRERE, M. H.   1975.   Control of Water Pollution from Cropland. Vols. I and II. U.S. Dept. Agric. and Environ. Protection Agency, Washington, DC.

USDA.   1978.   A Bibliography for Small and Organic Farmers, 1920 to 1978. U.S. Dept. Agric., Washington, DC.

USDA.   1980.   Report and Recommendations on Organic Farming. U.S. Dept. Agric., Washington, DC.

VAIL, D. and ROZYNE, M.   1978.   Appropriate technology for small farms: Lessons from 31 organic farms in Maine. Idea Paper *11*, Main Small Farm Management and Technology Project. Dept. Econ., Bowdoin College, Brunswick, ME.

WISCHMEIER, W. H. and SMITH, D. D.   1978.   Predicting Rainfall Erosion Losses: A Guide to Conservation Planning. USDA Agric. Handbook *537*.

3

Crop Energy Conservation

Today the food-producing capacity of American agriculture is unsurpassed by that of any foreign power. The efforts of farmers and agricultural scientists over the last 40 years are responsible for this great achievement. Since 1950 these efforts have resulted in a doubling of productivity and a halving of agricultural labor (Price 1981). Much of the increase can be attributed to increased use of fertilizers and advances in pest control, plant and animal breeding, and mechanization. With the exception of results based upon plant and animal genetics, most of the efforts that increased productivity were energy dependent. The price of doubled agricultural productivity is high: a four-fold increase of energy input (Price 1981).

Estimates place the energy consumption of the food and fiber system at just over 20% of the total 1981 energy needs in the United States (Price 1981; Vilstrup 1981). The agricultural segment of the food and fiber system requires at most one-sixth of this system's energy: The remainder is needed for food processing and distribution (Fig. 3.1). The latter includes energy to transport food from the farm to the processor and eventually to the supermarket, and energy to process, package, market, store, and prepare the food.

Figure 3.1 is based on 1980 data. In comparison with 1978 data, energy consumption for field machinery, crop drying, and miscellaneous activities remained about the same in 1980. However, a decrease of 34 and 8.8% occurred for irrigation and transportation, respectively, and an increase of 14% for fertilizers and pesticides, due mainly to increased fertilizer use. The decrease in energy usage attributed to transportation probably resulted from conservation measures and improved fuel efficiency of newer vehicles. The decrease in energy consumption for irrigation probably resulted largely from improvements in irrigation efficiency and some loss of irrigated acreage to dryland farming, a result of energy costs becoming too high for profitable operation of irrigated land in some areas.

At first glance Fig. 3.1 appears to show that crops consume most of the agricultural energy input. However, a significant part of the corn, alfalfa, and soybeans is fed to livestock. Animal husbandry, including operations and feed requirements, is thus the main energy consumer in American agri-

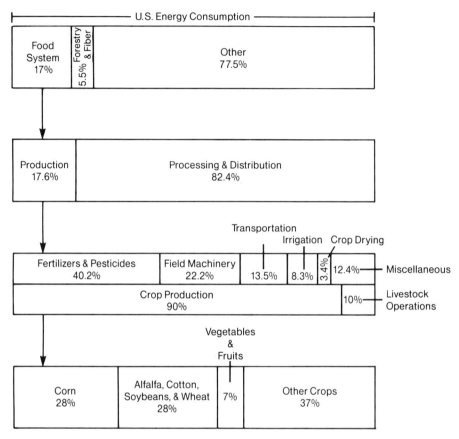

FIG. 3.1. Flow chart of U.S. energy consumption indicates relative energy usage in various components of the agricultural production system based on 1980 data.

culture. This has led some to question whether livestock production should have such a substantial role in American agriculture.

The energy value of crops before they are processed or fed to livestock exceeds the energy input required for their production. This favorable energy balance changes considerably once the energy input for food processing and distribution is added, or the energy value of livestock products is weighed against the direct consumption of feeds used to raise livestock. This negative energy deficit has been accepted in the past, since it rested on inexpensive and readily available energy and produced high-quality food products that were easily available and convenient.

Recently, some analysts have begun to examine agriculture strictly on an energy basis. All food and fiber products are thought of in terms of energy output and input, as was done in the preceding paragraph. Others look at agricultural data strictly in the economic sense. The cost of energy to produce American agricultural products on the farm is about 10% of their value (Lockeretz 1982). In this view, the food needs of America are satisfied at a reasonable cost to the consumer, and excess food remains for export. Indeed, the value of agricultural exports amount to somewhat over half the cost of imported oil, yet their production on the farm only consumes approximately 1% of our national energy budget. This group feels less alarm about energy input to agriculture and gives it only that attention that economic trends foretell.

Still others think that agricultural products may someday reduce our need to import energy. They point to ethanol from corn, diesel-like fuel from sunflower oil, and petroleum feedstocks from certain oil-rich succulents cultivated on arid lands. Here two factors must be watched: energy input vs. energy output and competition between crops for food or energy.

Regardless of which viewpoint one takes, energy use is a common consideration in all cases. As such, energy conservation is an integral part of sustainable agriculture. However, the holistic view of sustainable agriculture should be kept in mind: Energy conservation must be accorded equal status with conservation of soil and water resources. Resource conservation must also be balanced against maintenance of maximal agricultural productivity with minimal labor input, while avoiding environmental pollution.

Energetics of Crop Production

An examination of the literature shows there is much variation in the energy needed to produce various crops. Factors involved in the wide range of energy inputs include the type of crop, the region, and the manner in which it is produced. Price (1981), for example, reported the following energy requirements, expressed in Btu per pound: tobacco, 30,000; feed grains, 2000; cotton, 24,000; and corn, 1200–8400. The energy input for corn depends on the intensity of the production system and the region. (For conversion purposes, 1000 Btu/lb equals 2.32 MJ/kg.)

These numbers must be put into proper perspective for purposes of evaluating the need for energy conservation in agriculture. Most studies estimate that energy used in food production on the farm accounts for about 3% of total national energy consumption; the highest cited value in the literature is near 5%. Energy consumption for crop production is rather low compared to that for other major activities, such as food processing and distribution at about 17% (Price 1981; Vilstrup 1981).

However, the dependence of crop production upon energy is much more critical than numbers would appear to indicate. A shortfall of energy at a critical junction, such as during periods of pesticide application or during crop harvest, could spell disaster on a national scale. Timeliness of the fuel supply is extremely important. Unfortunately, the timing of fuel requirements is itself a variable that depends upon factors such as crop development, weather, and appearance of pests.

One obvious way to reduce the potential danger of temporary energy shortfalls and to cut production costs is to decrease the total energy input. However, before this can be done, several questions require answers. What crops are energy intensive? Are the energy-demanding crops a significant part of agricultural production, such that energy conservation would produce a tangible result? How is energy partitioned during crop production; that is, could one activity in crop production be made more energy efficient? Are some cropping systems more energy efficient than others? These and other questions can only be answered after an energy audit is conducted for the various crops.

Agricultural scientists have conducted energy audits for the major U.S. crops (Pimentel *et al.* 1973; Heichel 1978). These audits are essentially an accounting of the partitioning of energy resources in crop production. Some of the components of these audits reflect direct energy inputs; these include the fuel needed to run farm structures and various machinery. Other components reflect indirect energy inputs, for example, the energy needed to make fertilizers, various agricultural chemicals, farm machinery, and buildings.

An excellent source of indirect and direct energy values for crop production is available (Pimentel 1980). Some examples of interest follow. The total embodied, fabrication and repair parts energy during the reliable life spans of a two-wheel-drive 130-hp tractor (6078 kg or 13,400 lb) and a 4.27-m (14-ft) chisel plow (2213 kg or 5100 lb) is 99,655,341 and 42,696,767 kcal, respectively. If the life spans of the tractor and plow are 12 and 15 years, respectively, and they are used on a 200-ha (494-acre) farm, the annual combined energy cost for both is 55,755 kcal/ha (22,573 kcal/acre). Energy inputs for farm-delivered fertilizers range from 12 to 15 Mcal/kg N, 0.6 to 1.5 Mcal/kgP_2O_5, and 0.5 Mcal/kgK_2O. The energy input for crushed and ground limestone is 300 kcal/kg. The average energy input for pesticides is 61,470–86,910 kcal/kg active ingredient for insecticides; 62,770–99,910 kcal/kg for herbicides; and 27,770–64,910 kcal/kg for fungicides.

These and other energy inputs for crop production have been summarized for each crop, and then the values proportionately adjusted to compensate for differences in days to maturity from seed or transplants. The final value gives an indication of the average daily rate of energy use per acre in producing a specific crop. The units of this fossil energy flux (FEF) are Mcal/acre-day. Values of FEF for the major crops and their relative importance in the consumption of agricultural energy are shown in Table 3.1.

Some interesting conclusions can be drawn from Table 3.1. Certain crops require much less energy for production than others. The least energy-

Table 3.1.   Acreage and Fossil Energy Flux of Major Crops in the United States

Crop	Acreage[a] (1000 acres)	Fossil energy flux[b] (Mcal/acre-day)	Relative energy rank
Corn			
silage	9,261	17	1[d]
grain	70,061	17	
Soybeans	67,856	5	2
Alfalfa	26,269	6	6
Wheat (spring)	19,479	5	7
Cotton (lint)	13,214	14	5
Sorghum (grain)	12,722	16	4
Oats	8,640	4	10
Barley	7,233	6	9
Vegetables and fruit	6,044.8	56[c]	3
Rice	3,295	34	6
Peanuts (for nuts)	1,398.8	18	11
Sugar beets	1,187.2	11	12
Sugar cane	735.6	9	13
Tobacco (flue-cured)	511.8	125	8
Pasture			
fertilized	—	2	—
fertilized, irrigated	—	23	—
Range	—	0.03	—

[a]USDA (1982).
[b]Heichel (1977, 1978); Klopatek (1978); USDA/FEA (1976).
[c]Average value.
[d]Combined silage and grain.

intensive crop is native range, used by grazing animals as forage. The only management is occasional applications of herbicide to control brush. Small amounts of energy are used to manufacture and apply the herbicide. At the other extreme tobacco requires nearly 4200 times as much energy as native range. Much of this energy consumption occurs in the curing and drying required shortly after harvest. Heavy use of liquified petroleum gas is needed to produce flue-cured tobacco. Fortunately, the acreage devoted to tobacco is small compared to most crops; little potential exists for energy conservation during the flue-curing process, except by eliminating tobacco production.

Vegetables and fruits have intensive crop systems involving heavy inputs of fertilizer, pest control, and irrigation. As expected, their FEF value is quite high (56), nearly 1900 times greater than that of native range. The input of fertilizer to convert native range to fertilized pasture having managed forage species raises energy consumption 66-fold; the addition of fertilizer and irrigation causes a 767-fold increase! The energy price for fertilizer, irrigation, and pest control is quite high.

An insight into one way of altering a cropping system to reduce energy consumption can be gained from looking at the values for corn and alfalfa. Silage corn is an annual forage requiring heavy nitrogen fertilization. Alfalfa is also an annual forage, but being a legume, it fixes about 80% of its nitrogen needs from the atmosphere. As shown in Table 3.1, the FEF of silage corn is 183% greater than that of alfalfa. Heichel (1978) has shown that the different needs for nitrogen fertilizer accounts for much of the energy differential between corn silage and alfalfa. Since leguminous forages after harvest leave nitrogen in the soil for succeeding crops, they can reduce nitrogen fertilization for nonleguminous crops. Before the ready availability of chemical fertilizers, farmers used legumes and nonlegumes in rotational crop systems to reduce the need for nitrogen. An old idea has found merit again. The use of this type of rotation is examined later in this chapter in the section on fertilizers.

Another approach to energy conservation in crop production is suggested by a comparative study of organic and conventional farms located in the Corn Belt (Lockeretz 1978). Each organic–conventional pair was matched as closely as possible in terms of soil and size. The organic farmers used only organic fertilizers (livestock manure, leguminous green manures, or commercial organic fertilizers) and natural forms of pest control; the conventional farmers used conventional techniques, chemical fertilizers, and pesticides. The organic operations produced 11% less market value per hectare, but needed only 40% as much fossil energy per hectare as conventional operations. Much of the energy consumption of conventional farming was attributed to the use of nitrogen fertilizers. Efforts to conserve energy by use of organic fertilizers are examined later in this chapter.

The studies of Heichel (1978) and Lockeretz (1978, 1982) clearly show that reducing the use of chemical forms of nitrogen fertilizer can conserve energy. However, their research also pointed out other aspects of agricultural production that consumed considerable energy: farm machinery, tillage, fuel, and pesticides. In decreasing order the three most energy-intensive and clearly defined crop production activities are fertilization, field operations (farm machinery and tillage), and irrigation (Heichel 1978; Waldrop 1981; Lockeretz 1982). Energy considerations associated with these and other crop production activities are examined in detail in the rest of this chapter.

Fertilization

A third of the productive capacity of American crops results from the use of nitrogen fertilizers. Estimates are that the United States would require 18 million acres of additional cropland to maintain current production levels if nitrogen fertilizers were restricted to 50 lb/acre (56.8 kg/ha) (Price 1981).

The production and application of fertilizers accounts for an estimated 30–38% of the total energy input for crop production (Lockeretz 1982; Price 1981). Fuel energy is needed for mining, refining, transporting, and applying fertilizers. Most of the inherent energy of fertilizers is attributed to the natural gas needed to produce nitrogen fertilizers. Average values of inherent energy in fertilizers (not including energy of field application) are 25,000 Btu/lb of nitrogen; 3000 Btu/lb of P_2O_5 equivalent; and 2000 Btu/lb of K_2O equivalent (Frye and Phillips 1981). In terms of gallons of diesel fuel equivalent per pound, the respective values are 0.170, 0.020, and 0.014; in terms of megajoules per kilogram (MJ/kg), the values are 58, 7, and 4.6, respectively.

Natural gas supplies the hydrogen needed for ammonia compounds, of which 94% are manufactured from gas. About 93% of the nitrogen in fertilizers is derived from ammonia compounds (anhydrous ammonia, urea, and ammonium nitrate). Anhydrous ammonia is slightly more energy efficient to manufacture than urea and ammonia nitrate. The respective average energy requirements are 46, 69, and 65 MJ/kg. Nitrogen fertilizers consume about 2% of the total natural gas used in the United States (Fluck and Baird 1980).

Fertilization application rates vary considerably depending on soil conditions, climate, and crop. Rates on corn, for example, range from 75 to 175 lb/acre (85 to 199 kg/ha) of nitrogen and from 60 to 150 lb/acre (68 to 170 kg/ha) each of potassium and phosphorus as P_2O_5 and K_2O. Nitrogen is the most important nutrient for production of nonleguminous crops because it leaches the easiest and tends to be deficient, thus limiting yields more than potassium or phosphorus. Nitrogen fertilizer, being used the most and also the most energy expensive, would seem amenable to more efficient utilization to save energy.

Reduction of Fertilizer Usage

Whatever energy conservation measures are taken to reduce usage of nitrogen fertilizer must be carefully considered against yield. The general relationship of yield to applied nitrogen fertilizer is shown in Fig. 3.2. The rate of increase in the yield of corn (or of any other crop) diminishes per additional increment of fertilizer as the peak yield is approached. However, this still may be economical, if the value of the additional corn is reasonably greater than the cost of the fertilizer. Conversely, any reduction in applied fertilizer, which results in lower levels of utilizable nitrogen, has a price tag: reduced yield. This need not be the case, if the reduction eliminates nonutilizable nitrogen normally lost by leaching or if nitrogen fertilizer is replaced by another less energy-expensive source of nitrogen, such as manure or legumes.

The manufacture of fertilizers could be improved in terms of energy use and ability to use lower-grade feedstocks. Some progress is being made on

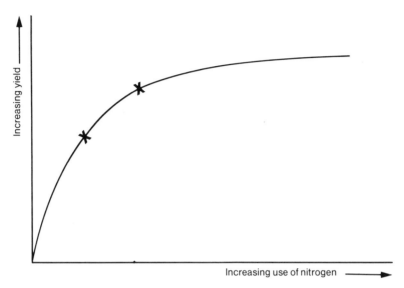

FIG. 3.2. Application rates of nitrogen fertilizer are usually chosen to realize yields in the range indicated by the Xs. Efforts to reduce nitrogen use should not be at the expense of yield, but aimed at eliminating excesses or practices that lead to lost nutrients.

this front with phosphate nutrients, which are closer than other nutrients to exhausting feedstocks. High-grade Florida phosphate rock is being used up at an alarming rate. Fortunately, research on process improvement and production of phosphate fertilizer from lower-quality ores has been started by the Tennessee Valley Authority's National Fertilizer Development Center (Anon. 1983).

Some approaches to fertilization of cropland can result in decreased usage of chemical fertilizers without causing reductions in crop yield. The amount of fertilizer applied should be based on thorough soil tests prior to initial and supplementary applications of fertilizer. Recommendations of these tests should be closely followed to avoid using too much (or too little) fertilizer.

However, even fertilization rates based on soil test results may be excessive (Liebhardt 1983). Results of soil tests are made on the basis of the best available information, which often is weak in terms of sufficient field calibration data. Also, good tests for available soil nitrogen are lacking, and results appear to be highly variable among laboratories. Many do not take into consideration the amount of nitrogen contributed by a previous legume crop. Often, even when soils test high in fertility, additional fertilizer is recommended. Evidence is beginning to suggest that money is now spent on some fertilizer that will not increase crop yields (Olson *et al.* 1982).

Improvement of soil testing, in terms of chemical tests and the data basis upon which results are formulated, appears to be needed. Such improvement, based on present evidence, warrants high research priority. The energy and monetary savings possible through reduced fertilizer usage based on better soil testing could be substantial (Liebhardt 1983).

A good review of the problems with the existing tests for predicting the release of soil nitrogen is available (Barker 1980). Some progress has been made in improving tests for the determination of nitrogen supplied by the soil, for example the method of Fox and Piekielek (1978a). While their test has been shown to be useful under several environmental and soil conditions (Fox and Piekielek 1978b), much research remains. Tests, such as this one, need to be calibrated to crop demands in the field under a wide range of environmental and soil conditions. Although this research will be expensive and time-consuming, the rising costs of nitrogen fertilizer provide considerably justification for the effort.

Careful attention should be given to the particular nitrogen fertilizer used and the timing and placement of application (Barker 1980). For example, fall-applied nitrogen is only 37% as effective as side-dressed nitrogen applied when corn is 30–45 cm (12–16 in.) high. Side-dressings are more efficient than spring preplant broadcast treatments. Urea-based fertilizers are more apt to lose nitrogen through volatilization than other types if no rain falls within 3 or more days after application. These same fertilizers also have been shown to produce lower yields than ammonium nitrate with no-till corn. Research is needed to reduce losses associated with urea-based fertilizers.

If fertilizer is applied prior to planting, some leaching or denitrification of nitrogen may occur. Such preplant treatments are therefore somewhat inefficient. For example, corn, which consumes 25% of the total energy consumed in U.S. crop production, has only a modest nitrogen requirement during the first 4 weeks after seed is planted (Frye and Phillips 1981). If the application of nitrogen fertilizer were delayed for 4 weeks, it would be reasonable to expect more efficient utilization of nitrogen. This can be the case, if climatic and soil conditions are correct.

Nitrate leaching is the most likely cause of nitrogen loss on well-drained soils. In soils of moderate to poor drainage, greater losses of nitrogen more likely occur through denitrification of nitrate. Increased rainfall accelerates both leaching and denitrification on soils, regardless of drainage conditions. Conventional tillage, as opposed to no-tillage, also increases losses from denitrification and leaching of nitrate.

When nitrogen fertilizers are applied all at planting, the rates for moderately well-drained to poorly drained soil are 50 and 75 lb/acre (57 and 85 kg/ha), respectively, higher than the rate on well-drained soils. This additional nitrogen is to correct for greater potential losses on soils with less than well-drained conditions. These rates can be reduced by 35 lb/acre (40 kg/ha) if up to two-thirds of the nitrogen fertilizer are applied 4 weeks after planting the corn (Frye and Phillips 1981). Such delayed applications appear to be most

Table 3.2. Time of Peak Nitrogen Requirements of Certain Crops

	Peak nitrogen demand	
Crop	Days[a]	Stage
Field corn	30–45	—
Sweet corn	25	Whorl
Cucumber	—	Fruit development
Lima beans	—	Flowering and fruiting
Radish	—	Root formation until 75% complete
Southern peas	—	Peak vegetative growth, pod filling
Soybeans	—	Peak vegetative growth, pod filling

Source: Based upon data in Barker and Mills (1980).
[a]After planting.

efficient for no-tillage corn grown on moderately well-drained soil and for conventional tillage corn on moderately to poorly drained soil. Periods of peak nitrogen requirements for various crops are shown in Table 3.2. Depending upon future research, crops other than corn might also be amenable to delayed applications of nitrogen fertilizers, with associated reductions in total usage.

Use of Organic Wastes

Other approaches to reduction of the energy costs involved with use of chemical nitrogen fertilizers include partial to complete substitution by legumes and/or organic wastes. The latter includes manures, crop residues, compost, and sewage sludge. Current USDA estimates are that 90% of the available manures are already applied to land. Even if all the unavailable manures (range droppings) were utilized, manure could supply at most one-third of the nitrogen needed for the total U.S. corn crop (Fluck and Baird 1980). However, substitution of manure for chemical fertilizers on corn does result in an energy savings of up to 40% (Lockeretz 1978). Substitution with manures works best when animals and crops are raised in close proximity. The cost of transporting manure more than 20 miles exceeds its fertilizer value. Generally, the use of manures to conserve energy and to improve levels of organic matter is most often feasible on organic farms.

Manure applications are not without problems. Special application machinery is needed, and improper applications can lead to surface runoff carrying nutrients, oxygen-demanding materials, and infectious agents into waterways. Excessive use of manure can cause nitrate pollution of the groundwater, increased soil salinity, and nutrient imbalances leading to poor crop growth. Excellent guidelines do exist for the proper utilization of animal wastes on cropland and pastureland; these are discussed in Chapter 7.

In conclusion, manures, which comprise about 22% of the organic wastes in the United States, have been recognized already for their value in reducing nitrogen fertilizer usage, and hence agricultural energy consumption, and in increasing soil organic matter. But, as noted already, an estimated 90% of the annual manure production (175 million dry tons) is already applied on land. A limitation appears to have been reached for this form of energy conservation. Other organic wastes have some nutrient value, but their main use is to maintain organic matter and to control erosion. These wastes are discussed in Chapter 7.

Use of Legumes — *saves energy*

Another very old agricultural method for supplying nitrogen, as old as the use of manures, is the utilization of legumes. However, this practice fell into disfavor with the advent of chemical fertilizers. Recently, interest in legumes have been renewed in view of the need to supply nitrogen economically and energy efficiently. Much of this interest centers upon the use of legumes as winter cover crops and in rotations. The amount of nitrogen fixed per acre by various legumes is shown in Table 3.3.

Table 3.3. *Annual Nitrogen Fixation by Legumes*

Crop	Nitrogen fixed[a]	
	(kg/ha)	(lb/acre)
Perennial		
Alfalfa (*Medicago sativa*)	193–220	172–196
Birdsfoot trefoil (*Lotus corniculatus*)	58–116	52–103
Clover:		
Landino (*Trifolium repens* f. *lodigense*)	200	178
Red (*T. pratense*)	83–149	74–133
White (*T. repens*)	115	102
Kudzu (*Pueraria lobata*)	120	107
Annual		
Bean (edible, *Phaseolus*)	45	40
Lespedeza (*L. cuneata, striata, stipulacea*)	95	85
Pea		
Cow (*Vigna unguiculata*)	100	89
Garden (*Pisum sativum*)	80	71
Winter (*Lathyrus hirsutus*)	60	53
Peanut (*Arachis hyogaea*)	47	42
Soybean (*Glycine max*)	65–110	58–98
Vetch (*Vicia*)	67–112	60–100
Biennial		
Sweetclover (*Melilotus*)	130	116

[a]Values taken from numerous sources.

Research indicates that legumes can supply substantial nitrogen to crops that follow. One very promising method is the use of legumes with no-tillage corn. Instead of using the residue of the preceding corn crop to prepare the mulch cover, a legume winter cover crop is used. The benefits of this approach are several, but our interest at this point is the energy conservation achieved through the use of less fertilizer.

Results with this approach have been encouraging. In one study, corn was grown in a no-tillage system with a winter cover crop of hairy vetch. No nitrogen fertilizer was used, but corn yields were equivalent to those observed with 99 kg/ha (87lb/acre) of nitrogen. With mixed red clover and hairy vetch as the cover crop, corn yields were equivalent to those obtained with 112 kg/ha (100 lb/acre) of nitrogen. Several other reports cite similar values for hairy vetch. Big flower vetch also seems to give similar results (Frye and Phillips 1981; Frye et al. 1983).

Top yields of corn are possible with legumes supplemented by modest amounts of nitrogen fertilizer. The agronomic value of this practice is clear. Its economic value, while presently borderline in terms of legume seed versus fertilizer costs, will become increasingly positive as prices of fertilizer escalate. In addition use of legume cover crops provide important erosion control benefits (Frye et al. 1983; Palada et al. 1983).

One estimate of the value of legumes in terms of energy conservation has been made by Frye et al. (1983). Legumes, if used on one-half of the no-till corn acreage, would conserve more than 50 billion megajoules of energy or 8.5 million barrels of petroleum annually by the year 2000.

Legumes can also be used in rotations with conventional crops in order to conserve energy. Numerous rotational schemes with corn are possible with annual and perennial legumes, especially those that are useful as forage crops. Some examples follow, but details are given in Chapter 7. Soybeans, as annual legume, can be alternated on a 1-year basis with 2 years of corn. This rotation reduces energy needs by 26% compared with continuous corn (Heichel 1978). It must be realized that soybeans, like other annual legumes, do little to reduce the nitrogen needs for the corn that follows. Soybeans receive about 30% of their nitrogen from the atmosphere and about 70% of this is removed with the harvested crop.

Perennial legumes (see Table 3.3) generally have better rates of nitrogen fixation and, hence, the potential to leave nitrogen for crops that follow. These perennial legumes can be used for forage in a rotational scheme with crops such as corn, wheat, and soybeans. The yield of the food crops may be reduced through loss of space to the forage legume, but increasing prices and demand for forage crops as cattle feeds are making such rotations increasingly cost effective (Heichel 1978; Lockeretz 1983). However, an alternative is to overseed legumes with corn and not lose space to legume rotations. The legume continues growth after harvest through early spring and is then plowed under. A new planting of corn and legumes follows. Early research shows that yields appear similar to straight corn; fertilizer needs and erosion are reduced (Palada *et al.* 1983).

The legume of choice for forage purposes and energy reduction appears to be alfalfa, as it leaves substantial amounts of nitrogen in the soil for succeeding grains. After one harvest, the nitrogen left in the roots and unharvested regrowth of a 2- or 3-year-old stand of alfalfa is equivalent to 190 lb/acre (213 kg/ha) of nitrogen. Proper use of alfalfa in rotational schemes with food crops may reduce nitrogen needs up to 50% (Heichel 1978). Alfalfa used in rotations with corn can reduce energy input compared with continuous corn by 39%. Rotations with corn, soybeans, wheat, and alfalfa can reduce energy input over continuous corn by 44%. These energy savings are attributed in the largest part to reduced usage of nitrogen fertilizer, but some savings are due to reduced need for fuel and other inputs (Heichel 1978). Triplett *et al.* (1979), based upon their research, suggest that a good meadow stand of vigorous alfalfa can supply nearly all the required nitrogen for a succeeding grain crop, whether no-tillage or conventional tillage is used.

The value of perennial legumes in conserving energy can undoubtedly be improved through plant breeding. Some improvements in the nitrogen-fixing ability of alfalfa has been achieved in breeding programs (Heichel *et al.* 1981a; Barnes *et al.* 1981), but much work remains to be done.

Much research also remains in terms of our understanding of legumes. The present methods used to assess nitrogen contributions from legumes appear to be questionable (Heichel and Barnes 1984). For example, they do not account for incomplete recovery of nitrogen by succeeding crops, nor are they capable of differentiating between normal soil-derived or symbiotically fixed nitrogen. It also appears that rotations improve yields for reasons in addition to increased nitrogen; yield increases have been observed in rotations involving only nonlegumes versus those with legumes and nonlegumes. Present methods fail to differentiate between benefits from symbiotically fixed nitrogen and the other poorly defined cofactors. Still Heichel and Barnes (1984) indicate forage legumes could replace 25–50% of nitrogen fertilizer needs of high-yield cropping systems.

We also need to better understand the effect of different management practices upon nitrogen contributions from legumes. For example, an unharvested legume plowed under would have a greater nitrogen benefit than the residues from harvested forage legumes. However, the amount in the residue that becomes available is a function of the time of soil incorporation, the amount removed for forage, and the frequency of removal (Heichel *et al.* 1981b). As Heichel and Barnes (1984) have pointed out, the constraints imposed by various management practices upon the substitution of biologically fixed nitrogen for chemical fertilizers need to be clearly expressed.

The problems notwithstanding, the importance of legumes to energy conservation is significant. As a final perspective, if one-quarter of the nation's corn crop were grown in an annual rotation with alfalfa, the annual savings in nitrogen fertilizer would free up 70 billion cubic feet of natural gas.

Field Operations: Farm Machinery

As mentioned already, the energy consumption associated with the use of farm machinery includes direct and indirect components. The indirect component reflects the energy inherent in construction materials and the energy used to manufacture and repair farm machinery. The farmer has little control over this manufacturing energy, and efforts to conserve such energy depend on market forces and the availability and cost of alternative materials and manufacturing processes. However, a program of energy conservation is possible for the direct component of energy consumption by farm machinery. Most of this direct energy usage is related to fuel consumed during tillage, planting, pest control, and harvest.

Fuel used in farm machinery represents a significant part of the energy consumed in crop production; operation of farm machinery consumes on an average about 20% of agricultural energy (Lockeretz 1982). For example, fuel accounts for 30% and 42% of the FEF for corn silage and alfalfa, respectively (Table 3.1). Decreased or more efficient utilization of farm machinery could be a reasonable avenue for energy conservation. One possible way to decrease fuel usage is through less tillage, which is explored in the next section.

The type of fuel used—gasoline, LP gas, or diesel—has a bearing on energy consumption. In 1960 most farm machinery was fueled by gasoline and LP gas. Today, most new farm machinery is diesel powered. Diesel engines convert fuel more efficiently to mechanical energy, and diesel fuel is both the least expensive and most energy rich per gallon. These advantages far outweigh the slightly higher prices of diesel engines over others. When you consider that fuel and oil to operate farm machinery costs U.S. farmers near $5 billion per year, a powerful economic stimulus exists for fuel conservation.

Energy also can be saved through proper selection, operation, and maintenance of farm machinery. Equipment most amenable to energy reduction appears to be the tractor and tillage attachments. Comparative test data on the performance of tractors is available from the Department of Agricultural Engineering, University of Nebraska (Lincoln, NE 68503), which operates a tractor-testing facility. Results from tractor tests can be used in a manner similar to that of purchasing a new car. For example, recent tests show considerable differences (23%) in fuel economy at maximal power takeoff. A number of factors have been shown to affect fuel economy: weight to power ratio, engine efficiency, number of drive wheels, weight distribution, and soil characteristics. Selection of implements for primary tillage is also critical. A moldboard plow produces a clean tilled seedbed, but uses 50% more energy than a chisel plow and 70% more than a heavy offset disk (Bloome *et al.* 1981). If a seedbed with small amounts of crop residue can be tolerated, considerable energy savings are possible. Implement width is also important in tractor selection. The type and width of implements determine what horespower is required at the tractor's drawbar output. Several factors determine implement width: speed of movement in the field, timeliness, and field efficiency. These factors all require careful balancing for most efficient energy usage and economic return.

Field efficiency is expressed as (actual work rate)/(theoretical work rate). Values normally range from 70 to 85%. Time spent on nonproductive operations—turning, making adjustments, overlap between runs, and repairing breakdowns—reduces efficiency below 100%. The higher the speed of tillage and the wider the implement width (up to a point), the greater the actual work rate. This results in increased productivity and reduced labor intensity. The increased horsepower needed per acre to increase productivity and reduce labor does have a cost: The initial investment in equipment is higher. This must be weighed against the economic advantage of increased productivity. Energy savings are therefore possible through downrating of tractor horsepower. However, this must be weighed against timeliness, which accounts for much of the increased yield associated with increases in horsepower. Timeliness is essentially the ability to conduct an activity during such a time as to maximize product quality and quantity. Timeliness is critical at certain periods: planting, harvesting, and sometimes pest control. Yields can be reduced 5–10% if operations are carried out only a few days from the optimal time.

In general, the tractor with the highest fuel economy should be selected. However, it should also have sufficient speed and drawbar horsepower output to handle the implement width that allows timely operations at the most critical periods. This would seem to call for overpowered and oversized equipment, but it is justified in terms of increased yield.

It must also be remembered that during those periods when timeliness is not so critical, fuel energy needs are less and the tractor can be run at optimal fuel economy. This fuel usage improvement is possible because diesel engines are most economical when operated at two-thirds of their rated power and rated speed. Tests of this concept, popularly called "gear up and throttle down," have demonstrated average fuel savings of 27% compared to full throttle runs (Fluck and Baird 1980; Bloome *et al.* 1981). Care must be taken to avoid overloading the engine.

Future tractors will undoubtedly have small computerized monitors that show energy consumption as affected by numerous variables: throttle setting, gear changes, weight distribution (ballasting), tillage depth, and so on. The operator will be able to set conditions to maximize fuel economy for most operations, but to maximize work rate during critical periods.

An important aspect of saving energy and money is proper maintenance of farm machinery. Good maintenance insures that equipment will run efficiently, last longer, and go for longer periods of time between repairs. Engine maintenance is of special importance; the best fuel economy comes from a well-tuned engine and clean air filters. Proper maintenance can increase fuel efficiency by up to 25% (Bloome *et al.* 1981).

Other factors can contribute to more efficient usage of fuel. Unnecessary idling should be eliminated: If you take a break, turn it off. Whenever possible, field operations should be combined so as to reduce the number of passes through a field. For example, the formation of a seedbed, fertilization, and sowing of seed might be done in one pass. Tillage reduction, to be discussed next, might be considered. Finally, the use of tractor weight for ballast is critical. Too much weight causes soil compaction and greater use of fuel, while too little means increased tire wear and wheel slippage.

Field Operations: Tillage

Tillage, or the working of soil for plowing, planting, and cultivation, is a significant part of agriculture. Estimates are that tillage operations move enough soil in the United States per year to create a superhighway from Los Angeles to New York (Frye and Phillips 1981).

Tillage appears to have peaked in the 1970s. Increased energy costs have encouraged farmers to consider no-tillage and reduced tillage systems. The energy requirements for conventional tillage (moldboard plow), reduced tillage (chisel plow, disk), and no-tillage systems are shown in Table 3.4. Although a no-tillage systems uses only about half the fuel required by conventional tillage, it must be remembered that tillage uses far less energy in crop production than fertilization. Switching from conventional to limited tillage would result in a 3% reduction in the energy requirements for production of corn (Fluck and Baird 1980); switching to no-tillage corn would give a 7% reduction (Phillips *et al.* 1980). In contrast, the use of legumes on a rotational basis with corn can reduce energy consumption about 10 and four times greater than switching to limited and no-tillage, respectively (Heichel 1978).

However, tillage reduction is being utilized not only to conserve energy but more importantly to reduce soil erosion and, to a lesser degree, increase water filtration and conservation. Since the loss of topsoil is perhaps the most

Table 3.4. Estimated Energy Requirements for Several Field Operations and Inputs in Four Tillage Systems

Input or operation	Gallons diesel fuel/acre			
	Conventional tillage	Chisel plow	Disk	No-tillage
Moldboard plow	1.84			
Chisel plow		1.12		
Disk	0.63	0.63	0.63	
Apply herbicides and disk second time	0.73	0.73	0.73	
Spray herbicides				0.13
Plant	0.43	0.43	0.43	0.50
Cultivate (once)	0.42	0.42	0.42	
Herbicides	1.75	2.01	2.25	2.88
Machinery and repair	1.86	1.61	1.25	0.60
Total	7.66	6.95	5.71	4.11

Source: Frye and Phillips (1981).

serious problem in agriculture (Sampson 1982), the move to reduced tillage is important. The scarcity of water supplies in some parts of the United States is a further impetus to adoption of reduced tillage.

With reduced tillage (also called minimum or conservation tillage) crop residues are incorporated nearer the surface, since the chisel plow and the disk produce more shallow tillage than the conventional moldboard plow. This acts to increase water filtration and to reduce soil erosion. Reduced tillage requires less mechanical power than conventional (less depth of soil to move), and hence uses less fuel (Table 3.4).

No-tillage systems are even more effective than reduced tillage systems for conservation of soil, water, and energy. However, the use of herbicides is necessary to control weeds normally eliminated by cultivation, and increased amounts of other pesticides often are needed to control insects harbored in the large amounts of crop residue. Even counting the energy required to apply pesticides, a no-tillage system is the most energy efficient (Table 3.4). While reduced tillage is acceptable to both conventional and organic farmers, no-tillage systems are not acceptable to organic farmers because they require use of pesticides.

Other advantages of reduced and no-tillage have been noted (Fluck and Baird 1980; Frye and Phillips 1981). Higher yields are possible, especially with well-drained land and dryland farming; reductions of labor and time also occur. In some instances double-cropping potential has been advanced northward by several hundred miles as a result of intensification of land use.

Sometimes soil structure is improved. Lower-quality land (e.g., with slopes up to 20%) may be farmed for row crops, and machinery investment may be decreased with reduced or no-tillage systems.

Besides increased usage of pesticides, other disadvantages of reduced and no-tillage systems have been noted (Fluck and Baird 1980; Frye and Phillips 1981). Some soils with fine textures produce lower yields when farmed with reduced tillage than when farmed with conventional tillage. More nitrogen fertilizer is needed for increased yields with no-tillage for some crops, such as corn. This increase of nitrogen means a greater potential for water pollution through nitrate leaching. Fortunately, the increased yield of corn more than compensates for the cost and energy value of the additional fertilizer. Increased problems from insects and rodents have been noted. Larger amounts of seed are required.

It would appear that the advantages of no-tillage and reduced tillage outweigh the disadvantages. At present about 3% of U.S. cropland is farmed with no-tillage; most of the no-tillage crop is corn. Reduced tillage is used for a wide variety of crops on about 25% of U.S. cropland. States that lead in no-tillage are Indiana, Kentucky, Missouri, and Pennsylvania. Leading states for reduced tillage include Illinois, Iowa, Kansas and Nebraska.

Irrigation

About 18% of U.S. cropland is irrigated and quite productive; such acreage produces about 27% of our crops (Jensen and Kruse 1981). In the 1980s, irrigation accounted for 13% of the total energy used in agricultural production (Price 1981). More recent data indicates a drop to 8.3% (Fig. 3.1). Irrigation ranks third as an energy-consuming component of crop production (transportation is partly associated with livestock operations). However, if only irrigated land is considered, the relative energy input associated with irrigation rises substantially. As energy costs have risen, the rapid expansion of irrigated farmland has slowed because the added costs started to counterbalance the increased productivity of irrigated land.

The amount of energy needed for irrigation is a function of several variables, the most important of which are pumping lift, required water pressure, frequency of irrigation, the type of irrigation system, and the net depth of the applied water. Pumping lift is the vertical distance between the water source and the discharge point over which the water must be pumped. Increasing the water pressure requires additional energy; a water pressure of 670 kilopascals (100 lb/in.^{2}) is equivalent to lifting water 70.4 m (230 ft).

Lower water pressures of 170–345 kilopascals (kPa) are used to operate sprinkler systems with hand-move and side-roll sprinkler laterals. Highest pressures (over 860 kPa) are needed for traveling and stationary hydraulic gun sprinklers. Moderate pressures of 520–690 kPa are used on center pivot spinklers. In surface irrigation systems dependent on gravity flow, less energy is needed than with sprinkler systems that require pumping. Another system with minimal energy requirements is trickle irrigation, where low water pressure is needed. Assuming a water lift of 100 m, the annual energy input for these systems would be about 4000 and 8700 kcal/ha, respectively, for surface irrigation (with an irrigation runoff recovery system) and traveling hydraulic gun sprinklers (Batty et al. 1975).

A close look at our largest crop, corn, can be used to put energy requirements into perspective. In the High Plains of Texas 60% of the crop production energy for corn is attributed to irrigation; this figure becomes 85% with nitrogen fertilizer added. In the Corn Belt, the energy input is about two-fifths of this value, mainly because irrigation is not used. Energy input for irrigation can be even higher. If pumping is required from deep groundwater aquifers, as in Arizona, irrigation can consume 90% of the crop input energy (Jensen and Kruse 1981).

A number of possibilities for conserving irrigation energy (and water) exist (Stetson et al. 1975; Heermann et al. 1976; Eisenhauer and Fishbach 1977; Ross 1978; White 1978; Jensen and Kruse 1981). Several of these are discussed in the following sections.

Reduction of Pumping Frequency

Reducing irrigation pumping frequency to the minimum can save energy. For example, although irrigation at certain times has been proven unnecessary for some crops, based on historical precedent, it is still done. In Idaho sugar beets that are kept well irrigated through August 1, such that the soil has 200 mm (8 in.) of available water, require no additional irrigation until harvest. The last application makes harvest easier. Once grain has reached the soft dough stage, additional irrigation seems to be of little value.

Care must be taken not to eliminate or reduce irrigation during critical stages of development. The critical stage for most grain crops and corn is the flowering and silking stage, respectively. The heading stage on small grains and flowering stage of soybeans is also critical. Irrigation at these times can return the greatest yield per unit of water. For example, average yields over 3 years with unirrigated corn and corn irrigated at the early silking stage differed by 2820–2950 kg/ha (45–47 bu/acre) in Kansas.

Pump Efficiency

Pumps must be operated at peak efficiencies through proper maintenance. New pumps with the greatest efficiency should replace aging, inefficient equipment. An electrical pump's efficiency is that of the pump, whereas the efficiency of pumps powered by natural gas or diesel fuel is a function of the combined efficiency of the engine and the pump. Optimal efficiency appears to be near 75%. Based on pump efficiency tests, one can determine the approximate reduction in energy use that can be achieved by improving efficiency to the maximal limit. The equation for the potential percentage reduction in annual energy use and costs is $(1-Ee/Ea)$ 100, where the existing efficiency is Ee and the attainable efficiency is Ea.

The potential reduction can be used to ascertain the feasibility of steps needed to raise efficiency. Sometimes this is simply a case of adjustment, maintenance, or parts replacement. At worst it may involve replacement of the entire pump. These steps involve variable capital costs, which must be weighed against the potential reduction of energy costs.

Improved Management

Better management of irrigation schedules can also result in reduction of energy consumption. The best management practice is to employ soil moisture detection devices coupled with computerized scheduling of irrigation. One study in Nebraska showed that corn yields were not greatly reduced when pumped irrigation water was cut up to about 50% over the season. In another study use of computerized irrigation schedules for center pivot sprinklers resulted in a 27% reduction of energy usage, with no effect on corn yields.

Estimates are that energy savings of 30–50% for water pumping are possible through improvements in management practices and equipment efficiency. Adjustments alone on many pumping plants could improve efficiency and reduce energy consumption by 13%.

System Alterations

Other possibilities for reducing irrigation energy requirements include alteration of application methods or shifting to other procedures such that energy intensiveness is reduced. One alteration is the conversion of a high-pressure sprinkler system to a lower pressure, such that the total dynamic head is reduced (TDH = pumping lift + pressure head + friction). This conversion requries more than changing nozzles; pumping plant modifications are also required (Jensen and Kruse 1981). Such changes are possible on center pivot systems, but the costs versus energy savings savings require careful evaluation. If a high-pressure center pivot system was of such an age as to necessitate replacement, a newer low-pressure center pivot system would be an ideal energy-saving investment.

As discussed earlier surface irrigation systems are more energy efficient than sprinklers, especially if the surface system is well designed. The distribution requires leveled land, channels of closed pipe or lined but open conduits, and a runoff reuse system. The latter is a small reservoir to trap surface runoff, which is returned to the irrigation system. Since the pumping lift is much less for surface water than for groundwater, energy costs are minimal. The energy requirements for pumping surface water are 15–20% lower than those for pumping groundwater.

In some situations sprinklers could be replaced by a surface system. Some surface systems can be made more energy efficient simply by adding a reuse system. Studies in Nebraska indicate that the addition of reuse systems can reduce pumping costs by 25%, since the recycled runoff reduces the time that the groundwater pumps are operated (Jensen and Kruse 1981). Another more energy-efficient replacement for sprinklers is a trickle system.

The use of gravity pressure rather than pumping lift can cut energy consumption. This method is possible by diversion of a mountain stream to a pipeline leading to sprinklers in valleys. Gravity flow of water for surface irrigation as opposed to pumping can also reduce costs. Upstream river diversion at points higher than the surface-irrigated area is required for this approach.

Improvement of irrigation efficiency can also cut energy consumption. Techniques include more uniform application of water, nighttime applications to reduce evaporation, application of only the amount of water retainable in the root zone, reduction of surface runoff, and elimination of leaks and seepage. A most-promising energy-saving innovation is the photovoltaic-powered water pump. Tax credits are available for their installation. Functional units are now in use in the United States (Arco 1982) in areas far removed from utility lines.

Pesticides

In terms of the crop production energy budget, pesticides are not nearly as demanding as fertilizers. Fertilizers compose 98% of farm chemicals; pesticides only 2%. For example, pesticides utilized in corn production consume about 0.8–3.5% of the energy input (Pimentel *et al.* 1973; Heichel 1978). The production and application of pesticides, according to the Council for Agricultural Science and Technology (CAST 1977), consumes about one-seventh as much energy as fertilizers. Thus, the elimination of pesticides would not result in a significant conservation of agricultural energy.

However, despite only modest potential savings in energy, there appear to be many compelling reasons for judicious use of pesticides and integrated pest management. Problems with pesticides include development of resistance among pests exposed continuously to pesticides and deterioration of environmental quality.

A number of practices can be used to control pests without pesticide use. These include weed control through tillage, mulches, sanitation, closer crop spacing, crop rotation, and proper timing of seeding and planting. A large number of natural controls are possible for suppression of insects and disease. These practices are discussed in detail in Chapter 9.

Crop Drying

Annual energy consumption for the drying of crops (Fig. 3.3) is at least 0.111×10^{12} megajoules (Fluck and Baird 1980). The practice of removing moisture from crops (e.g., corn, soybeans, rice, peanuts, and tobacco) with hot air derived from burning fuels, mostly natural or LP gas, has increasingly replaced natural air drying. Crop drying accounts for 3.4% of the total energy consumed during crop production (Fig. 3.1). Dried crops are easier to store because they are less likely to spoil, overheat, or become infested with insects.

At present about half the U.S. corn crop is dried directly on farms and another quarter off the farm. Drying, besides aiding good storage, permits earlier harvesting, the growing of higher-yielding, later-maturing cultivars, and, in some areas, double cropping. These advantages incur an energy

FIG. 3.3. Crop-drying facilities are dependent primarily upon LP and natural gas. Such fuels are prime candidates for replacement by solar energy.
Courtesy U.S. Department of Agriculture

debt. For example, the fuel required to dry 1 ton of corn from 25–15% moisture is 18.4 liters of LP gas. Expressed in other terms, the same 10% reduction in moisture for the average yield of corn in a hectare (2.47 acres) would require 156–196 liters (39 to 49 gal.) of LP gas. Some alternative would appear desirable, especially in view of the gas shortages in the Midwest during the recent decade, which curtailed the drying of corn.

Some energy-saving alternatives do exist (Frye and Phillips 1981). These include field drying, low-temperature drying, solar drying, dryeration, and direct preservation of high-moisture grain destined for cattle or hog consumption (see Chapter 5).

One of the simplest and most promising techniques, if the advantages of crop drying have high priority, is dryeration. Grain is removed hot from the drier just short of acceptable moisture contents and slowly aerated-cooled after being held for 4–12 hr with aeration. It may also be possible to dry grain at lower temperatures and slower rates in certain regions. This procedure is feasible for harvests in cool and not overly humid regions, but not in areas like the southeastern United States. In these hotter, more humid areas, it is more feasible to dry first at conventional high temperatures until the moisture is reduced enough to allow storage lengths acceptable for a second drying stage, such as prolonged low-temperature drying or natural air drying.

Delayed harvest results in field drying. However, the savings in energy may be negated by possible yield decreases and additional field losses. Present late-maturing cultivars are often not suitable for delayed harvest, and earlier cultivars frequently yield less. Damp, rainy weather may pose problems in terms of disease and drying. In some areas the potential of a double crop might be lost. One possibility, if planting conditions permit, is earlier planting of later-maturing cultivars, such that harvest occurs at the usual time but with drier grain (Morey et al. 1978). Field drying requires careful assessment to insure that yield losses are not equal to or greater than energy savings.

Solar energy appears to be promising for grain drying (see Chapter 10), and current tax advantages for solar equipment help to offset the high initial expense. Because grain drying is done only during a short period of time, solar systems must be inexpensive or serve multiple purposes (Spillman 1981). The future looks bright for slow low-temperature solar drying, but not fast high-temperature drying (Chau and Baird 1978). Another promising technology is heat pump driers (Peart et al. 1980).

Frost and Cold Protection

Protection of frost- and chilling-sensitive crops consumes energy, usually in the form of burning fuel and movement of air by wind machines (Fig. 3.4). The annual energy expenditure is about equivalent to that for greenhouse

FIG. 3.4. **Frost protection in citrus groves uses energy to run wind machines and heaters, as shown here.**
Courtesy American Society for Horticultural Science

heating, 0.045 $\times$ 10^{12} megajoules (Fluck and Baird 1980). Frost and cold protection ranks far below crop drying as an agricultural energy consumer, but its low rank can be deceptive. For example, in some years citrus groves require no frost protection, whereas in other years energy needs for such protection may be 60.8% of the required energy in citrus groves (USDA/FEA 1977). Energy consumption for a typical hectare in a citrus orchard might range from 400 to 1600 liters (105.6 to 422.4 gal.)/hr of oil depending on ambient temperature. The benefits of energy conservation are obvious in these conditions.

Conservation techniques include an assessment of the energy costs for producing the crop versus the energy costs for frost protection. If the weather abnormality persists, it could be economic disaster to continue protection. At some point saving the crop becomes unjustified, as energy costs for producing a new crop become so much less than the energy expended for cold protection. Other possibilities are the use of sprinklers for minor cold protection instead of heaters and wind machines. For example, annual energy input for frost protection of cherries with oil heaters is 84,708,760 kcal/ha but is only 8,161,000 kcal/ha with irrigation (Pimentel 1980). Sprinklers can also be used to delay blossoming through evaporative cooling, thus bypassing unexpected late frosts.

A long-range solution may be found in existing research. Under some conditions water may be supercooled in liquid form to $-5°C$; under these conditions, frost protection could be withheld more so than is the acceptable practice. Since most frost protection efforts are attempted in the temperature range of $0°$ to $-5°C$ range, much of the energy need for cold protection could be eliminated. However, supercooling is prevented if nucleation centers for the formation of ice crystals are present. Nucleation centers are provided by bacteria associated with soil and plants, such as *Erwina herbicola* and *Pseudomonas syringae*. Plants themselves have no such nucleation centers. Scientists have isolated a bacterial gene that is involved with the nucleation center (Anon. 1982a). Eventual research may lead to an inhibitor or alterations through genetic engineering, thus reducing energy costs for cold protection. One such inhibitor, a polymer, appears close to reality (Anon. 1982b). Kocid 101, a fungicide and bactericide, became licensed for light frost protection through control of ice-nucleating bacteria in 1983.

Transportation

Transportation ranks third as an energy consumer in agricultural production (Fig. 3.1). Its rank in crop production is somewhat difficult to determine, since transportation also utilizes considerable energy in livestock production. Based on the available data, it is estimated that roughly one-fourth of the transportation activity is associated with animal production. This estimate would mean that transportation associated with crop production uses about the same, or slightly more, energy than irrigation. The relative rank (whether third or fourth) of transportation and irrigation in energy consumption in crop production cannot be accurately determined with the present data.

Attempts to cut energy usage in farm transportation depend presently upon conservation. Vehicles must be properly tuned and lubricated if optimal fuel economy is to be realized. Transportation patterns should be as efficient as possible. Two or three trips with light loads should be replaced with one trip. Several errands over a week might possibly be combined into one trip. As aging vehicles are replaced, newer ones should be as fuel efficient as possible.

In the long run, farm transportation can only become more sustainable if alternate fuels from renewable resources become available. For example, ethanol can be produced through fermentation of grains, such as corn. Ethanol can be mixed with gasoline (gasohol) or used alone to run gasoline-dependent vehicles. Some adjustments are required, but they are possible. A few farmers, how many is unknown, run pickup trucks and cars on ethanol now. Ethanol can also partly supplement diesel fuel, but considerable modifications are involved.

However, widespread adoption of alternate fuels for gasoline and diesel is in the future. A discussion of alternate fuels including present technology, needed improvements, and constraints is presented in Chapter 10.

Future Directions

The present approach to reducing energy usage in crop production focuses on conservation. As long as energy supplies are sufficient and conservation measures compensate for price increases, this would seem to be a prudent choice. However, as energy supplies inevitably dwindle in the future, agriculture must eventually move toward energy sustainability. The most logical approach to achieving this is use of alternate energy sources derived from renewable resources.

The prospects of reaching that goal are highly mixed. For example, the biggest energy consumer, nitrogen fertilizer, could become sustainable first through the use of legumes and then with crops having genetically induced biological nitrogen fixation. The use of legumes is presently feasible, and indeed was once the common practice. With more research, the widespread adoption of this practice is not insurmountable if the need arises for an interim or partial solution. The ideal way to reduce usage of nitrogen fertilizer is the genetic engineering of our present crops to meet their nitrogen needs like legumes, that is, through biological nitrogen fixation. Here the prospect is far less promising, until considerable research is accomplished. The major breakthrough needed to realize this goal is not imminent, but more likely far in the future (see Chapter 10).

The impetus for achieving nitrogen sustainability is the present energy intensiveness of chemical fertilizers plus the availability of a simple renewable resource, atmospheric nitrogen, that could replace them. The push for sustainability in terms of phosphate and potash is not as hard, since energy consumption and soil needs for these two nutrients are considerably less. However, the ore resource for phosphate is now seen as limited. Sustainability of these nutrients probably could be realized through recycling efforts, for example, by applications of unprocessed or composted organic materials (see Chapter 7). However, it is not now clear what extent nutrient needs could be satisfied through this approach, nor what would constitute optimal management practices.

Direct use of energy on the farm can also become a sustainable practice. Fuel use accounts for most of the energy consumption associated with field operations and transportation. Farmers presently use some 6 billion gal. of fuel annually, about equally divided between gasoline and diesel. A significant impact upon energy sustainability would result if these were replaced with alternate renewable fuels. The basic technology for doing so is presently available, although improvements and more incentives are needed if widespread adoption is to take place. A careful analysis is also required to assess the effects of producing alternate fuels in substantial quantities on food prices, food supplies, and soil erosion. Some improvement in the compatibility of combustion engines to alternate fuels is also required.

The alternate fuel that presently shows the most promise is alcohol, especially ethanol. The technology for producing ethanol is available and already being used, although to a very limited extent. The fermentation process that produces ethanol also produces animal feed as a byproduct. Ethanol can be used to supplement or replace gasoline with minimal difficulty; it also can supplement diesel fuel although this requires substantial engine modification. Alternately, diesel tractors could be replaced by tractors with different engines that utilize straight ethanol. Such units exist in Europe. A diesel fuel equivalent can also be prepared from vegetable oils, but substantial research is needed to overcome problems associated with this technology. An indepth discussion of alternate fuels is presented in Chapter 10 on future technology.

Another alternate energy source is solar energy. The basic technology is available, although refinements and some new technology are still needed. The most likely scenario is partial, gradual adoption of solar energy in the near to moderate future, especially for grain drying, hot water heating, and space heating. Solar devices for these purposes have already been adopted by a few users. See Chapter 10 for additional information on solar energy.

The adoption of the preceding approaches to sustainability will be mixed. Which practice or practices and to what extent each will be utilized depends upon not only national needs, but regional and even individual farm characteristics. The adoption of sustainable practices and technologies will be influenced by several factors: farm size, crop mix, livestock presence, conventional energy availability and price, cost of alternate systems, management skills, do-it-yourself ability, and information availability. The future pattern of sustainable crop production will not be uniform, but a mosaic consisting of mixed practices that best meet local needs.

References

ANON. 1982a. Gene may protect crops from frost damage. Chem. & Eng. News (Aug. 2), 6.
ANON. 1982b. Polymers may be able to block frost in crops. Chem. & Eng. News (Oct. 4), 31–32.

ANON.   1983.   TVA set sights on future phosphate technology. Chem. & Eng. News (Nov. 7), 37–38.

ARCO   1982.   Photovoltaic-powered water pumps growing in popularity. ARCO Solar News 2 (4) 4–9.

BARKER, A. V.   1980.   Efficient use of nitrogen on cropland in the Northeast. Conn. Agric. Exp. Stn. Bull. 792. New Haven, CT.

BARKER, A. V. and MILLS, H. A.   1980.   Ammonium and nitrate nutrition of horticultural crops. Hort. Rev. 2, 395–423.

BARNES, D. K., HEICHEL, G. H., VANCE, C. P., VIANDS, D. R. and HARDARSON, G. 1981.   Successes and failures while breeding for enhanced nitrogen fixation in alfalfa. In Genetic Engineering of Symbiotic Nitrogen Fixation. J. M. Lyons, R. C. Valentine, D. A. Philips, D. W. Rains and R. C. Huffaker (Editors). Plenum Publishing, New York.

BATTY, J., HAMAD, S. N. and KELLER, J.   1975.   Energy Inputs to Irrigation. J. Irr. and Drainage Div., ASCE 101 (IR4), 293–307.

BLOOME, P. D., GREVIS-JAMES, I. W., JONES, L. K. and BATCHELDER, D. G.   1981.   Farm machinery ideas that save energy. In Cutting Energy Costs (The 1980 Yearbook of Agriculture). J. Hayes (Editor). U.S. Dept. Agric., Washington, DC.

CAST   1977.   Energy Use in Agriculture. Rept. No. 68. Council for Agric. Sci. and Technol., Dept. of Agronomy, Iowa State, Ames.

CHAU, K. V. and BAIRD, C. D.   1978.   Solar grain drying under hot and humid conditions. In Proc. 1978 Solar Grain Drying Conf., Purdue University, West Lafayette, IN.

EISENHAUER, D. E. and FISCHBACH, P. E.   1977.   Comparing costs of conventional and improved irrigation systems. Irrig. Age 11 (8) 36–37.

FLUCK, R. C. and BAIRD, C. D. 1980. Agricultural Energetics. AVI Publishing Co., Westport, CT.

FOX, R. H. and PIEKIELEK, W. P.   1978a.   A rapid method for estimating the nitrogen-supplying capability of soil. Soil Sci. Soc. Amer. J. 42, 751–753.

FOX, R. H. and PIEKIELEK, W. P.   1978b.   Field testing of several nitrogen availability indexes. Soil Sci. Soc. Amer. J. 42, 747–750.

FRYE, W. W. and PHILLIPS, S. H.   1981.   How to grow crops with less energy. In Cutting Energy Costs (The 1980 Yearbook of Agriculture). J. Hayes (Editor). U.S. Dept. Agric., Washington, DC.

FRYE, W. W., HERBEK, J. H. and BLEVINS, R. L.   1983.   Legume cover crops in production of no-tillage corn. In Environmentally Sound Agriculture. W. Lockeretz (Editor). Praeger Publishers, New York.

HEERMANN, D. F., HAISE, H. R. and MICKELSON, R. H.   1976.   Scheduling center pivot sprinkler irrigation systems for corn production in eastern Colorado. Trans. ASAE 19, 284–287.

HEICHEL, G. H.   1977.   Energy analyses of forage production. In Proc. Intern. Conf. on Energy Use Management, Vol. 1. R. A. Fazzolare and C. B. Smith (Editors). Pergamon Press, New York.

HEICHEL, G. H.   1978.   Stabilizing agricultural needs: Role of forages, rotations, and nitrogen fixation. J. Soil and Water Cons. 33, 279–282.

HEICHEL, G. H. and BARNES, D. K.   1984.   Opportunities for meeting crop nitrogen needs from symbiotic ntirogen fixation. In Organic Farming: Current Technology and Its Role in a Sustainable Agriculture. J. Power and D. Bezdicek (Editors). Amer. Soc. Agro., Madison, WI.

HEICHEL, G. H., VANCE, C. P. and BARNES, D. K.   1981a.   Evaluating elite alfalfa lines for N_2-fixation under field conditions. In Genetic Engineering of Symbiotic Nitrogen Fixation and Conservation of Fixed Nitrogen. J. M. Lyons, R. C. Valentine, D. A. Phillips, D. W. Rains, and R. C. Huffaker (Editors). Plenum Publishing Corp., New York.

HEICHEL, G. H., BARNES, K. D. and VANCE, C. P.   1981b.   Nitrogen fixation by forage legumes, and benefits to the cropping system. In Proc. 6th Annu. Symp. Minnesota Forage and Grassland Council. St. Paul, MN.

JENSEN, M. E. and KRUSE, E. G.   1981.   Cheaper ways to move irrigation water. *In* Cutting Energy Costs (The 1980 Yearbook of Agriculture). J. Hayes (Editor). U.S. Dept. Agric., Washington, DC.

KLOPATEK, J. M.   1978.   Energetics of managed land-use systems. *In* Proc. Intern. Conf. on Energy Use Management, Vol. 1. R. A. Fazzolare and C. B. Smith (Editors). Pergamon Press, New York.

LIEBHARDT, W. C.   1983.   Variability of fertilizer recommendations by soil testing laboratories in the Unites States. *In* Environmentally Sound Agriculture. W. Lockeretz (Editor). Praeger Publishers, New York.

LOCKERETZ, W.   1978.   Economic and resource comparison of field crop production on organic farms and farms using conventional fertilization and pest control methods in the midwestern United States. *In* Towards a Sustainable Agriculture. Jean-Marc Besson and Hardy Vogtmann (Editors). International Federation Organic Agriculture Movement Conference 1977. Verlag Wirz AG, Aarau, Switzerland.

LOCKERETZ, W.   1982.   Energy use by agriculture: How big a problem? *In* Proc. Cornucopia Project Symposium: Planning a Sustainable U.S. Food System, Lehigh Univ., 1981. Rodale Press, Emmaus, PA.

LOCKERETZ, W.   1983.   Energy in U.S. agricultural production. *In* Sustainable Food Systems. Dietrich Knorr (Editor). AVI Publishing Co., Westport, CT.

MOREY, R. V., GUSTAFSON, R. J., CLOUD, H. A. and WATER, K. L.   1978.   Energy requirements for high-low temperature drying. Trans. ASAE *21*, 562–567.

OLSON, R. A., FRANK, K. D., GRABOUSKI, P. H., REHM, G. W. and KNUDSEN, D. 1982.   Economic and agronomic impacts of varied philosophies of soil testing. Agron. J. *74*, 492–499.

PALADA, M. C., GANSER, S., HOFSTETTER, R., VOLAK, B. and CULIK, M. 1983.   Association of interseeded legume cover crops and annual row crops in year-round cropping systems. *In* Environmentally Sound Agriculture. W. Lockeretz (Editor). Praeger Publishers, New York.

PEART, R. M., BROOK, R. and OKOS, M. R.   1980.   Energy requirements for various methods of crop drying. *In* Handbook of Energy Utilization in Agriculture. D. Pimentel (Editor). CRC Press, Boca Raton, FL.

PHILLIPS, R. E., BLEVINS, R. L., THOMAS, G. W., FRYE, W. W., and PHILLIPS, S. H. 1980.   No-tillage agriculture. Science *208*, 1108–1113.

PIMENTEL, D.   (Editor)   1980.   Handbook of Energy Utilization in Agriculture. CRC Press, Boca Raton, FL.

PIMENTEL, D., HURD, L. E., BELLTOTI, A. C., FORSTER, M. J., OKA, I.N., SHOLES, O. D., and WHITMAN, R. J.   1973.   Food production and the energy crisis. Science *182*, 443–449.

PRICE, D. R.   1981.   Where farm energy goes. *In* Cutting Energy Costs (The 1980 Yearbook of Agriculture). J. Hayes (Editor). U.S. Dept. Agric., Washington, DC.

ROSS, R.   1978.   Colorado pump tests show how to make big dollars savings. Irrig. Age (March), 12(3) 9, 12, 16.

SAMPSON, N.   1982.   Soil, land use and abuse. *In* Proc. Cornucopia Project Symposium: Planning a Sustainable U.S. Food System, Lehigh Univ., 1981. Rodale Press, Emmaus, PA.

SPILLMAN, C. K.   1981.   Capturing and storing energy from the sun. *In* Cutting Energy Costs (The 1980 Yearbook of Agriculture). J. Hayes (Editor). U.S. Dept. Agric., Washington, DC.

STETSON, L. E., WATTS, D. G., COREY, F. C. and NELSON, I. D.   1975.   Irrigation system management for reducing peak electrical demands. Trans. ASAE *18*, 303–306, 311.

TRIPLETT, G. B. JR., HAGHIRI, F., Van DOREN, D. M.   1979.   Legumes supply nitrogen for no-tillage corn. Ohio Report (Nov.–Dec.), 83–85.

USDA.   1982.   Agricultural Statistics 1981. U.S. Dept. Agric., Washington, DC.

USDA/FEA.   1977.   A Guide to Energy Savings for the Orchard Grower. U.S. Dept. Agric., Washington, DC.

USDA/FEA.   1976.   Energy and U.S. Agriculture, 1974 Data Base. U.S. Dept. Agric. and Fed. Energy Adm., Washington, DC.

VILSTRUP, D.   1981.   Less energy, more food. *In* Cutting Energy Costs (The 1980 Yearbook of Agriculture). J. Hayes (Editor). U.S. Dept. Agric., Washington, DC.

WALDROP, M. M.   1981.   Deep changes taking root in U.S. agriculture. Chem. & Eng. News (June 1), 23–28.

WHITE, J. G.   1978.   Re-use pits: cheapest water on the farm. Irrig. Age (Nov.–Dec.), *12*(11), 58, 62.

4

Greenhouse Energy Conservation

Greenhouse agriculture is particularly vulnerable to problems of energy cost and disruption. Energy usage in greenhouses is concentrated, critical, and quite costly, whether growers operate in the North or South. For example, about 80% of the heating fuel in northeastern greenhouses is consumed in 5 months, November through March; 3 months, December through February, account for two-thirds of that usage. A disruption of energy on an icy cold day or extremely hot day could mean a disastrous loss of crops, or a sharp increase in energy costs could create serious cash flow problems. Fortunately, considerable energy savings are possible by modification of existing structures and by innovative designs of future facilities.

Because many of the existing greenhouses were constructed before the energy crisis, they were designed with the comforting knowledge that energy costs were not a major factor. Accordingly, such greenhouses are not as efficient in terms of energy conservation as greenhouses constructed in the past decade. Estimates are that less energy-efficient greenhouses (constructed before 1970) utilized in year-round intensive production cover about 2539 ha (6271 acres). This coverage does not include greenhouses used for seasonal production of transplants or other temporary usage.

The energy savings possible by modification of older greenhouses can be considerable. If night temperatures are maintained at 12.8°–18.3°C (55°–65°F), annual greenhouse fuel usage in northeast Ohio (Short and Bauerle 1981) is 15,336 dekaliters/ha (100,000 gallons/acre) of number 2 fuel oil or 160,567 m³/ha (14 million ft³/acre) of natural gas. Modifications, using present technology, could produce energy savings of up to 40%. At present prices of oil and gas, a grower could conceivably save up to $44,000 and $15,680 per acre in 1 year for oil and natural gas, respectively. These figures assume a boiler efficiency of 80%. This potential savings, of course, must be weighed against the payback period for the modifications (see later section in this chapter). In warmer areas heating costs would be less, but costs for summer ventilation would be proportionately higher. Achievement of savings is possible in several areas, each of which will be examined.

Fuel and Furnaces

The initial step in considering heating costs should be a careful analysis of fuel selection based on price. Fuel costs should be reduced to a common basis to facilitate comparisons. The ideal basis exists already in the fuel trade: cost per 100,000 British thermal units (BTUs). Comparative fuel costs, both present and projected, are shown in Fig. 4.1.

From these graphs an obvious conclusion is that natural gas and coal are

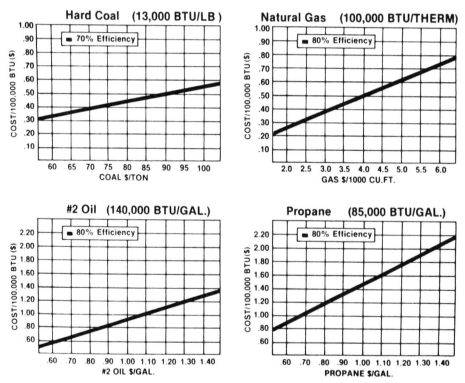

FIG. 4.1. Comparison of the costs of various fuels in terms of a common basis (cost/100,000 Btu) is useful in selecting fuels. Because actual fuel costs vary with time and region, the relative cost of various fuels is not constant. However, if one assumes that a grower has a choice of propane at 70¢/gal., fuel oil at $1/gal., and coal at $60/ton, the most economical fuel, based on these graphs, would be coal. The coal cost per 100,000 Btu would be about one-third that of propane and fuel oil.

From *Grower Talks*, ©1983, Geo. J. Ball, Inc., West Chicago, IL 60185

FIG. 4.2.   Incendo coal boiler has a reported efficiency of 80–84% and is claimed to be 25–33% less costly than comparable equipment. Coal feed is automated and ash removal semi-mechanized.
Courtesy Kewanee Boiler

much more economical fuels at present prices than oil or propane. Essentially, the costs for natural gas or coal are less than half that for oil or propane. If the existing furnace is near the age for replacement or grossly inefficient, its replacement by a natural gas or coal furnace makes economic sense. Even if not, the payback period may be such that it still is a reasonable option. This decision would be based on the age and efficiency of the boiler, prevailing interest rates, calculated energy savings, system costs, operational and maintenance costs, and projected lifetime of the new system.

Natural gas is not available everywhere, but coal nearly is. In fact, soft coals are available in some localities at prices below those of the hard coal used to calculate costs in Fig. 4.1. Those prices may make soft coals look attractive, even considering the extra furnace chores. However, chores are less with hard coal burners now available that heat from 0.25 to 2 or 3 acres. Their operation is reliable; no fireman is needed; the handling of coal, ashes and clinkers is automated; and pollution requirements can be met with these burners. Such units (Fig. 4.2) burn high-grade anthracite rice coal (¼ in.), which contains minimal sulfur and comes from Pennsylvania.

Reports are that at least 70 to 80 East Coast growers switched recently from oil to coal (Ball 1982a). Fully automatic coal furnaces are available in sizes from 7.5 to 500 hp to handle glasshouse areas ranging from 0.1 to 1.9 ha (0.25 to 4.6 acres). One report indicates that a grower with 0.6 ha (1.6 acres) under glass in Long Island, New York, switched with great success from oil to coal (Ball 1981a). This area is quite windy with a high wind/chill factor in the winter. Fuel costs were cut from $1.28 to 60¢/ft² and the whole installation was paid off in one year's fuel savings. A new coal boiler/burner has a reported efficiency of 80–84% (Ball 1983).

Wood and wood wastes, which are available in some areas at reasonable prices, may be attractive for some growers. These include sawdust from sawmills and wood product manufacturing firms, wood chips from arborist operations and land clearance/maintenance, and cordwood from lumbering activities. Values of fuel equivalents for various wood products are listed in Table 4.1. The fuel oil equivalency values do not take furnace efficiency into account. The costs of waste wood or cordwood can be compared with that of an equivalent amount of fuel oil. In many areas the cost of wood turns out to be 25–50% of the cost for an equivalent value of fuel oil. This must be weighed against the cost of wood heating systems and the increased costs for their operation and maintenance (Bartok 1981). Some precedence exists in the burning of wood for industrial use (Koch 1981).

If a new heating system is selected, the decision often comes down to a choice between a central boiler and unit heaters. In the past, unit heaters were the choice, especially for smaller greenhouse ranges, because of their flexibility and lower initial cost. If a new range were added or an older one expanded, then a new unit heater was added. Central boilers, boiler rooms, heating mains and return systems, and other accessories required a much bigger investment than several unit heaters.

Table 4.1.   Fuel Equivalencies of Wood Products

Product	Gallons of fuel oil
Sawdust[a]	
wet	50
dry	100
Cordwood[b]	
green	150
seasoned	240
Woodchips[a]	
fresh	50
aged, dry	100

[a]One ton.
[b]One cord.

Table 4.2.   *Typical Boiler Efficiencies*

Boiler type	% of CO_2	Exhaust temperature (°F)	Efficiency[a] (%)
Vertical, cross tube	8	1000	58
Unit heater	—	—	65–75
Lancashire	9	650	74
Vertical, firetube	10	700	75
Cast iron sectional	10	600	77
2-pass Scotch Marine	11	550	79
3-pass economic	$11\frac{1}{2}$	500	82
Economic packaged (3 pass)	13	450	85

[a]Of clean boilers; values can drop by as much as 20% in dirty boilers.

However, some disadvantages of unit heaters have become significant in view of energy problems, especially if the greenhouse range is larger than 0.2 ha (0.5 acre). Unit heaters are designed for operation with one fuel and are not easily converted. If a fuel supply is suddenly restricted—as natural gas was in Ohio during the winter of 1977 to 1978—big problems result for the greenhouse industry. Central boilers can easily convert from one fuel to another (e.g., from oil to natural gas or propane or even coal. If supplies drop or costs increase greatly for the current fuel of use, an economical alternative is feasible. Central boilers often have a better average fuel efficiency than unit heaters: 70–85% compared with 65%. Part of this efficiency difference is attributed to the modulation of boilers and their bulk as opposed to the on-off and small volume of unit heaters (Ball 1979).

Another important aspect of the heating system, either new or old, is its efficiency. The various types of boilers used to heat greenhouses vary considerably in their theoretical efficiency (Table 4.2). Moreover, a boiler may have the potential to operate at 80% efficiency, but run at 70% if it is dirty. Such a boiler, if rated to heat a greenhouse with 3200 gal. of fuel, actually needs 3657 gal. when dirty. It pays to select the most efficient boiler, but even more importantly it pays to keep it operating at maximal efficiency. Tests of boiler or furnace efficiency can be done with simple equipment in 10 min. A good explanation of the procedure is given by Bartok (1980). The return in energy conservation for a 10-min test can be substantial. Losses in efficiency are usually caused by dirty flues or incorrect fuel–air ratios.

Several other minor procedures can improve heating efficiency in a greenhouse. Thermometers should be checked for accuracy periodically, as heat could be expended needlessly if a thermometer is off (Botacchi 1980a). Thermostats should be accurate to ± 1°F and placed out of the sun, but not near the heat source. Air movement (use fans if necessary) should be maintained around the thermostat. Aspirated thermostats increase heating (and cooling) efficiency. If the greenhouse is subjected to prevailing winds, a windbreak could cut heating costs by 5–10% (Bartok 1977). Finally, painting heating pipes with aluminum or metallic paints, can reduce heat transfer by 15–20% (Maisano 1980). If paint must be used, latex paints are better.

An area of future interest in terms of greenhouse heating (and cooling) is the microprocessor, the future replacement of the thermostat. These computer components will be the future heart of very efficient greenhouse climate controls. Thermostats are somewhat inefficient in that temperature overrides occur after the thermostat stops calling for heat; microprocessors can anticipate and prevent temperature overrides, thus saving fuel. Temperature control is accurately maintained and constantly monitored. Expansion of ranges does not present a problem (Ball 1982b).

Microclimate Heating

The objective of microclimate heating is to heat the plants and not the air by directing energy to the plant roots and leaves. This can be accomplished by soil heating (Fig. 4.3) and/or infrared heating. The former warms the root zone and the latter the shoot. Such systems can save energy because heat is not expended on warming large air masses. Soil temperature is maintained at 18.3°–23.9°C (65°–75°F) and ambient air temperature at 7.2°–10°C (45°–50°F) instead of the usual 15.6°–18.3°C (60°–65°F).

Limited research has been done on microclimatic heating, but the results are promising. In one study on tomatoes, energy savings of 45% were achieved with soil heating, and 33–41% with infrared heating (Hughes 1981). White (1982) estimates that a double-layered greenhouse teamed up with microclimate heating, multiple-layered thermal blankets (see later section on insulation), and styrofoam insulation on bench undersides would need 90% less energy than a single-layered greenhouse with standard convection-air heat. One commercial greenhouse firm cut its fuel use by 83% by adding a thermal blanket and floor heating (Ball 1983).

The combined cost of soil heating and infrared heating would appear not to be justified for modern energy-efficient greenhouses with multiple-layer glazing and/or thermal blankets, but may be in old single-layer glass greenhouses (White 1982). Costs for installation are roughly $1.50–2.00 for infrared heating and 70¢ for soil heating per greenhouse square foot (Hughes 1981; Ball 1983). A brief summary of microclimatic equipment and grower evaluation was provided by White (1982) and Ball (1983). Future research will be needed to establish the optimal conditions of microclimate heating for each crop.

White (1982) has discussed some of the drawbacks of microclimate heating. Uneven heat distribution can occur with infrared heating of deep foliage

FIG. 4.3. Tubes under the pots in this root zone heating system carry hot water. Systems like this one are available commercially and are used both for propagation and the growing of plants. Root zone temperatures usually range from 65° to 75° F.
From Grower Talks, © 1982, Geo. J. Ball, Inc., West Chicago, IL 60185.

canopies, such as occur in carnations, cucumbers, large potted house plants, roses, and tomatoes. The foliage interception of heat is concentrated near the top, leaving the lower leaves and soil cooler (soil heating can help here). Problems have been experienced with some of these crops as a result of uneven heating. Infrared units operate at high temperatures, necessitating a height to spread factor of 1 : 2. Some greenhouses do not have adequate headroom and under these conditions older greenhouses are laid out such that considerable heat will be wasted on aisle space. Design work is needed to allow closer distribution of energy to the foliage and root zone. Soil heating requires careful engineering of each application for proper tube spacing, water temperature, water flow rate, and temperature sensor location. On very cold days air temperature may drop too low, but supplementary infrared heating would solve this problem. Snow loads must be closely watched in view of lower air temperature.

A promising breakthrough in soil heating is the adaptation of thin-film technology. A waterproof mylar sheet containing printed copper resistor strips and heat-generating film is now available. Temperature is easily controlled with a thermostat, and the system can be divided into four temperature zones. Heat is electrical, so no hot water is needed and bulkiness is thereby greatly reduced. Cost is reasonable, maintenance is essentially non-existent, and no extensive renovation is required for installation. The maker (Flexwatt Agritape, Avon Lake, Ohio) claims that these sheets can reduce overall energy costs up to 50%, with a cost payback time of 1 year. Increased growth and germination times were noted at air temperatures of 12.8°C (55°F) at night and bottom heat temperatures of 24.5°C (76°F). More research is needed on this promising new product.

Temperature Control

Two new concepts of temperature control, both designed to conserve energy without affecting plant productivity, have appeared in recent years. Both have evolved from the same premise that a minimal temperature can be held at night just short of a critical crossover point, after which plant development would be impaired. The critical period would be the time required at night for the translocation and metabolism of sugars produced by photosynthesis. Under natural conditions the night or day temperature is not constant, so it would be reasonable that some temperature variation is acceptable, rather than the set day/night temperature common to greenhouse operations.

One of these new approaches assumes that plants can be held at a minimal temperature part of the night, then allowed to experience a rise to a warmer temperature. This concept is termed the *split-night regime*. The other approach assumes that wide temperature ranges are possible, as long as the temperature average per day does not drop below some minimal temperature. This concept is termed *temperature averaging*. The split-night temperature regime saves energy by reducing heating units, whereas temperature averaging saves energy by allowing longer retention of sun heat, essentially a greater dependence upon solar energy.

Split-Night Regime

The idea of dropping the night temperature to a minimum for part of the night originated with New Haven, Connecticut, grower William Loefstedt (1977) through a series of informal experiments in the winter of 1977. Lilies, chrysanthemums, and geraniums were grown at 15.6°C (60°F) during the night until 11:00 p.m. and then dropped to 4.4°C (40°F) until one-half hour before sunrise at which time 15.6°C was restored. Daytime venting was at 23.9°C (75°F). At sunrise the plants were watered with warm water at 22.2°–23.9°C (72°–75°F).

The results were encouraging enough to interest scientists at the Connecticut Agricultural Experiment Station. Experiments by Station scientists

(Thorne and Jaynes 1977; Gent *et al.* 1979) and subsequently at the University of Connecticut (Schneider and Koths 1980; Koths and Schneider 1980, 1981) verified that the split-night regime saves energy for some crops. According to present research fuel savings of 15–20% appear possible. The concept is practical but does not work for all stages of growth for every greenhouse crop. Indeed, with some crops the treatment is beneficial for only certain cultivars (Parups and Butler 1982). Available results are summarized in Table 4.3.

The importance of watering with warm water shortly before sunrise needs to be resolved. In the studies of Gent *et al.* (1979) tomato and tobacco plants were watered at 7:00 A.M. with 50–250 ml of water at 21.2°C (70°F), whereas lilies were watered only once a week. It was noted that the soil temperature took 3 hr to reach that of the controls whose night temperature was not dropped. Development of the lilies was not retarded compared with that of the controls, suggesting that warm water might not be needed for lilies. Results of Koths and Schneider (1980) appear to indicate that there is no advantage to warming the soil of plants on split-night regimes. If warm water

Table 4.3.   *Crop Response to Split-Night Temperatures*

Crop	Split-night regime[a]	Developmental stage	Response[b]
Azalea		Reproductive	−
Begonia (Rieger)	A6, C6		0
Calceolaria	C6	Reproductive	+
Carnation	D6		−
Chrysanthemum	A6, B7, B8, C6	Reproductive	+
Cineraria	C6	Reproductive	−
Easter Lily	A6, B7, B8, C6	Reproductive	+[c]
Geranium	A6, B7	Reproductive	+, 0
Gloxinia	A6, C6		0
Kalanchoe			0
Marigold	B7	Seedling	+
Petunia	B7	Seedling	+
Poinsettia (early cultivars)	A6, C6		+
Snapdragon	D6		0
Tobacco	B8		−
Tomato	B8	Vegetative	+
		Reproductive	−[d]

[a]A = 15.6°/10.0°C (60°/50°F); B = 15.6°/7.2°C (60°/45°F); C = 15.6°/4.4°C (60°/40°F); D = 10.0°/4.4°C (50°/40°F). The number following the letter indicates the number of hours at which the minimum night temperature prevailed.
[b]+ = grows well under split-night temperature; − = slowed development; 0 = inconclusive results
[c]Some delay prior to bud formation.
[d]13% reduction in yield.

is not essential, some energy savings would result, but more importantly the split-night approach would have greater appeal to commercial growers. Since few crops have been tested, the unimportance of warm water for all crops cannot be assured.

Split-night regimes do save energy, and cost very little to implement. Essentially the only costs are the purchase and installation of two thermostats and a double-pole double-throw time clock. Certain crops succeed well under split-night temperatures (e.g., production of marigolds, petunias, and tomatoes as bedding plants and the forcing of pot chrysanthemums); some show slower development unitl their period of vernalization passes (e.g., Easter lily); and others show notable slowing of development (e.g., carnations and tobacco).

Temperature Averaging

The temperature-averaging approach had its origins at Cornell University, where it was established that some plants developed the same when the night temperature was constant and when the same temperature occurred as the average temperature for the night. The next logical step was to examine the effects of temperature averaging over both the day and night, with the hope that some fuel savings were possible. Such research is in progress now, but interesting preliminary results have been reported by Ball (1982b).

In a typical temperature-averaging regime, much of the daytime heat is retained and temperatures are allowed to reach 26.7°–29.4°C (80°–85°F). Venting during the day is minimal. At night the greenhouse is allowed to cool; no heat is used unless the temperature drops below 10°C (50°F). Thermal sheets are drawn at night across the greenhouse as an insulator to slow the loss of the heat built up from the trapping of solar energy during the day. Such an averaging scheme might be equivalent to a normal day/night range of 21.2°/18.3°C (70°/65°F), like that used for chrysanthemums. Results so far appear to be encouraging. Although no figures of energy savings are available, it appears likely that they would be at least comparable to those observed for the split-night temperature regime: 15–20%. Since a night reduction from 18.3° to 12.8°C (65° to 55°F) would result in a fuel savings of 26% when the average outside temperature is −2.7°C (27°F), estimated energy savings of 15–20% appear to be conservative.

The successful application of temperature averaging in commercial operations remains to be demonstrated. Temperature control is more complicated, and will certainly require the capability of microprocessors to maintain the ever-changing program of temperatures. A reliable thermal blanket will be required. Energy savings must be carefully quantitated, and optimal conditions for plant development for various crops under temperature averaging must be determined. Microprocessors are commercially available and are in successful use now; thermal blankets are already in use and demonstrate consistent energy conservation. However, the additional savings from temperature averaging remain to be verified. Optimal conditions are under study now at Cornell University.

Insulation

A number of insulating techniques and materials have been developed for use in older, less energy-efficient greenhouses. Some are conventional materials, such as polyurethane or Celotex boards for use in areas where light transmission is not important. Others are essentially light-transmitting materials for use overhead, where a drop in light transmission can be harmful to crop production. These forms of insulation vary in their efficiency, but fuel savings in the range of 20–40% would appear to be feasible.

Polyethylene Over Glass

The use of polyethylene over glass was developed at the Ohio Agricultural Experiment Station (Bauerle and Short 1977). Two sheets separated by an air layer (double polyethylene over glass, DPOG) were fastened over old glass greenhouses (Fig. 4.4). Coverage could include the north roof, all the roof (with or without vents included), and even the side walls. The more coverage, the greater the savings of energy. The air space was maintained by a squirrel-cage blower.

The insulation value of this system is unquestionable. The R value (resistance to heat flow) given in hr °F ft^2/BTU) for single glass is 0.88, and the R value for a double layer of polyethylene (clear, 2 to 6 mil) is 1.43. The two together (DPOG) have an R value of 2.00. This change in R value should translate into energy savings, and it does. However, the amount is dependent upon a number of variables: the condition of the glass area (leaky and loose or tight), the extent of the greenhouse covered, and the tightness of the installation.

The research from Ohio indicated fuel savings of up to 50%. Commercial growers report fuel savings from 33 to around 50% (Ball 1979). Such savings permit recovery of installation costs in a reasonable period of 1–2 years. However, other factors must be considered, as they may alter the economics of crop production such that the fuel savings may be illusory.

On the positive side, fuel savings are significant and no costly interior alterations ar needed. Interior purlin posts and the existing environmental systems pose no problems because the polyethylene is placed over the

FIG. 4.4. An air-inflated double plastic layer can be mounted over older, leaky glass greenhouses and provides an easy way to conserve energy.
From Grower Talks, © *1979, Geo. J. Ball, Inc., West Chicago, IL 60185*

exterior part of the roof. Temperatures hold steadier under marginal conditions when the heating system is taxed by extreme cold and high wind.

On the negative side are the following problems that can cut into profits. The most serious is an 18–20% reduction in light. This can lead to crop delays and loss of quality in areas where winter light intensities are close to being critical. Crops such as tomatoes, cucumbers, and roses may experience yield reductions of 5–10% (Short and Bauerle 1981) in a DPOG system. Polyethylene also deteriorates upon exposure, thus reducing light intensities even further in subsequent years. If the polyethylene is replaced annually, the deterioration problem is solved, but additional expenses are incurred. Using only a single layer of polyethylene reduces the amount of light lost, but energy savings will also be reduced. If light levels are raised by fluorescent or high-intensity discharge lamps, their cost must be considered. Other possible problems result from reduced air loss, including higher humidity, hence a possible increase in fungal diseases, and CO_2 depletion, which could lead to slower crop development.

If any of these factors lead to loss of crop quality and production delays, profits will be reduced. This profit loss must be weighed against the fuel savings possible with DPOG. Therefore, a careful cost analysis is required before adoption of DPOG. An excellent example of such a cost analysis was reported by Stefanis *et al.* (1981).

Bubble Pack

A bubble-pack plastic, similar to the type used in packaging operations, is available for greenhouses. Its R value varies between 1.69 and 1.87, depending upon the size of the air bubble. The material is attached to interior glass by capillary action, double-sided tape, or glue (Fig. 4.5). Bubble-pack plastic reduces light by about 12% and may cut heat loss by about 25%. At present it appears to be used primarily on hobby greenhouses, where its esthetic appeal wins over the appearance of DPOG. Caution is advised on overhead use, as snow accumulations are possible, leading to possible roof damage.

Thermal Sheets

A thermal sheet (heat sheet, thermal blanket) is a retractable layer(s) of insulating material that in winter is placed nightly gutter to gutter over the crops. The idea is to trap and slow the loss of heat in the crop area and not waste heat on overhead space, which also serves as an insulator. Such a sheet may be retracted by hand or mechanically (Fig. 4.6). Thermal sheets are made from various materials, such as plastics and synthetic fabrics, with R values ranging from 0.83 to 2.63 (Ball 1982a). Fuel savings appear to

FIG. 4.5. A bubble-pack plastic can be applied over the glass inside a greenhouse to save energy. One available material is AirCap®, shown here.
Courtesy Sealed Air Corporation

FIG. 4.6. Thermal blanket systems for relatively open greenhouses can serve as a thermal trap alone or as a combined thermal–shading unit. Good fuel savings and a reasonable cost payback period make these units excellent choices.
Courtesy Wadsworth Control Systems

range from 20 to 50%, depending on the material, condition of the green-house, and outside air temperature (Fig. 4.7). Average savings appear to be 33% (Ball 1982a; Short and Bauerle 1981), although double- and triple-layer thermal blankets are said to cut fuel by 50–75% (Ball 1983).

Installation costs are lowest and ease of operation greatest in greenhouses with minimal structural framework and with heaters and pipes below gutter height. If purlin posts are present, other insulation systems, such as DPOG, may be more economical. Installation and material costs vary from $0.30 to $2.75/ft^2 depending upon the material chosen for the thermal sheet (Ball 1982a). The cost payback period is 1–3 years, with tax-saving advantages taken into account.

A note of caution is that thermal sheets must fit snugly, or much of the fuel savings will be lost and water vapor seepage will condense and form frost on the roof, thus cutting light intensity. In general, there is little light loss, es-pecially compared with that associated with DPOG. After daytime retraction only a minor shadow is seen. Installation of thermal sheets is relatively per-manent; no annual replacement of sheeting is needed. Another advantage is the multi-functional nature of a thermal blanket; it can double as a day-length control system for short-day crops such as chrysanthemums and poinsettias or shade control system for shade-loving plants in the summer.

Thermal sheets may require minor maintenance, and with retractable sys-tems there is the risk of mechanical breakdown inherent in any moving system. Thermal sheets may also increase the risk of roof collapse during heavy snowfall because the energy efficiency of the sheets slows snow melt-ing. if the sheet is retracted during heavy snowfall, the problem is eliminated.

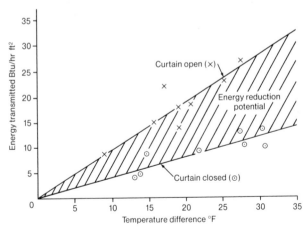

FIG. 4.7. Energy savings potential with thermal sheets are summarized in this graph. The greater the difference between the temperature inside and outside a greenhouse, the greater the energy savings when the thermal sheet is drawn.

From Rebuck, White, and Aldrich (1976) and the Pennsylvania Flower Growers' Bulletin No. 289

Some growers use a variation of the thermal sheet, a permanent sheet of polyethylene suspended gutter to gutter. Since it does not retract, considerable light loss may occur, and the danger of roof collapse from heavy snowload is possible. The advantage is cheapness of material and installation. Snow problems can be eliminated by installing a heating line about the sheet, which is only turned on during snowfall. Energy savings are possible with permanent sheets, but documentation is scarce; savings are probably comparable to those with retractable thermal sheets.

Thermal sheets offer a viable alternative to DPOG in older greenhouses of the clear-span type. They are also an essential part of the temperature-averaging system discussed earlier, and an invaluable adjunct to the microclimate-heating approach also discussed previously. Commercial systems are available, making thermal sheets an attractive energy saver (Ball 1979, 1982a, 1982b; Botacchi 1980b).

Sausage Tubes

Sausage tubes can be viewed as a product resulting from the combination of inflated double polyethylene (2 mil) with a thermal sheet (Fig. 4.8). Essentially, they are air-inflated single (sometimes double) polyethylene tubes

FIG. 4.8. Inflated polyethylene tubes, called sausage tubes, form an insulation barrier at night. During the day they are deflated and drawn to the side. They are a good choice as an inexpensive, short-term energy saver in greenhouses that may be closed down, renovated, or relocated within a few years.
From *Grower Talks*, © 1982, Geo. J. Ball, Inc., West Chicago, IL 60185

about 1–3 ft in diameter placed adjacent to each other and extending the length of the greenhouse. These form a thermal blanket, which can be drawn to the gutter edges when deflated (Ball 1982a). Their R value is similar to that of inflated double polyethylene. Sausage tubes produce energy savings comparable to those with DPOG and thermal sheets, about 35–40% (Ball 1982a).

The advantages of sausage tubes include no light loss during the day, heat retention at night, and the ability to fit around purlin posts, unlike thermal sheets. Since they are placed inside, deterioration of the polyethylene by ultraviolet light is negligible; hence, the cheapest grade of polyethylene can be used. Snow loads cannot damage them, as may occur with DPOG, and they can be deflated to melt snow on the roof. Payback time is on the order of 1 year. Some "bugs" remain to be worked out of the mechanical system for retracting sausage tubes, and hand drawing can become a burdensome chore. If they are left in place, some light loss results as with DPOG.

Lapseal

Overlap of glass on older greenhouses was somewhat loose, often leaving spaces or "laps" through which heat loss took place. A clear silicone-type sealant (Lapseal) can be injected between panes to close the laps. Lapseal can be applied with a hand or powered caulking gun. Fuel savings depend upon the looseness of glass and the outside temperature; savings of 5–50% have been reported, with the average being 15–25% (Short and Bauerle 1981; Ball 1979; Botacchi 1980b). Cost payback time is around 1.5 years.

Sealing glass laps produces fuel savings without annual reapplication of sealant (guaranteed for 10 years) and with essentially no light loss.Humidity levels tend to run higher in sealed greenhouses but pose no problem with proper venting and circulation. However, watering frequency is reduced. The one disadvantage of this technique is that the sealant makes it more difficult to replace damaged glass.

Miscellaneous Insulation

Various insulating materials (e.g., conventional fiberglass with vapor barrier and foil-backed styrofoam or urethane) can be applied to the north wall of a greenhouse with little light loss. Savings of fuel are estimated to be 10% (Salsedo 1980). Similar insulation can be applied to nonglass sidewalls and reflective aluminum insulation can be placed behind wall heat pipes, leading to fuel savings of 3–6%. The advantages are little or no light loss, especially with reflective insulation, and minimal expense.

Roofing Materials

Greenhouse construction today must be conducted with a major emphasis on energy efficiency. Various approaches are available, such as double-glass or inflated double-polyethylene structures. A number of new energy-efficient roof panels have also emerged. These panels, which can be used in new construction or to replace roofs on existing greenhouses, can cut fuel costs by 35–40%. These will be examined in view of their potential energy savings.

Acrylic Panels

A double panel made from acrylic is available as Exolite (Fig. 4.9). Each sheet is 1.5 mm thick and held apart by ribs spaced at 1.27 cm (0.5 in.), forming a rigid panel of 15-mm (0.6-in.) thickness. These sheets are secured by an extruded aluminum bar, producing a tight seal against air and water infiltration.

These panels have an R value of 1.79 and can result in fuel savings of 40–50% compared with glass. The weight of the panels with aluminum extrusions are comparable to that of glass; hence no extra structural support is needed.

One disadvantage of Exolite is cost: an estimated $5.25/ft^2 to recover an existing structure (Ball 1981b) and $7 for new structures (Ball 1983). More recent information indicates costs are coming down to an acceptable level, (Ball 1985). Light transmittance through the nonribbed section (80%) is somewhat less than through glass (90%); transmittance through ribbed sections is 75%. The longevity of Exolite and the effect of aging on light transmittance is not yet accurately known, as the product is relatively new. However, it appears to be stable for 20 plus years (Aimone 1985). One possible problem with Exolite panels is snow load buildup, necessitating either removal by hand or some form of gutter heat line to prevent light blockage and extra roof support. Finally, Exolite does burn and cannot be walked upon. Insurance costs might be higher, but its strength is not a problem under normal usage.

A limited number of growers are presently using Exolite in the United States; it is widely used in Germany (Ball 1981b). As cost becomes more competitive, more use of Exolite may be seen in the future.

Polycarbonate Panels

A number of double panels made from polycarbonate are available: Cyroflex, Qualex, Tuffak-Twin Wall, and Primex (Fig. 4.9). A polycarbonate version of Exolite is also available. Qualex is similar in construction to Exolite,

FIG. 4.9. Plastic ribbed panels are excellent for renovation and construction of greenhouses. *Top,* Exolite, made of acrylic; *bottom,* Cyroflex, made of polycarbonate. These and similar materials are good energy savers and transmit light well.

Courtesy CYRO Industries

but is thinner (7 mm, 0.28 inch). The two products are similar in light transmittance and fuel savings. The R value of Qualex is 1.72. Some discoloration occurs over time at a rate of 1% per year leading to a 10% loss at 10 years. If more loss occurs, the 10-year guarantee of no more than 10% discoloration is adequate backing. Snow load problems with Qualex are comparable to those with Exolite.

The cost of Qualex is about one-half that of Exolite; it also is one-third lighter than Exolite and thus requires less structural support. Because of these features, a Qualex installation costs about two-thirds as much as an Exolite one (Ball 1981b). Flammability is also minimal with Qualex, as it is self-extinguishing. Finally, Qualex is stronger than Exolite. At present only a few Qualex-covered greenhouses exist. Coatings to resist yellowing are being developed. If successful, this technology could push polycarbonates to the forefront of new greenhouse materials. One such recent product, Lexan Thermoclear Sheet, offers the hope of nonyellowing, but needs the test of time.

Tuffak-Twin Wall (R = 1.61) and Cyroflex (R = 1.45) are similar to Qualex. These have 80% light transmittance, impressive impact resistance, and a weight plus a price comparable to Qualex. These products will obviously be competitive with Qualex if polycarbonate double panels capture the market for greenhouse covers. At this point more experience is needed with polycarbonate panels.

Double-Glass Panels

A German double-layered glass, Sedo, is in wide use in Germany (Fig. 4.9), although little is currently used in the United States. Two glass layers are formed into a continuous panel containing a 9-mm (0.36-in.) air space. Air, moisture, and algae do not enter the air space. Carbon dioxide is used to fill the space, which avoids condensation problems.

Strength is excellent. On rare occasions heavy hail has broken the top panel layer, but not the bottom (Ball 1981b). Fuel savings with Sedo are comparable to those with Exolite and Qualex, and light transmittance is rated at 80–85%. Unlike the plastics, double glass does not discolor, has an almost permanent longevity, and poses no fire hazard. Two problems are noted, however. The snow load problem of double plastics is also present with double glass and the weight of double glass is such that supports must be twice as strong as with single glass. Cost is also high; an installation would run about \$8.50/ft^2 (Ball 1983).

Experience in Germany is good with Sedo. Hopefully, costs could be reduced by manufacturing the product in the United States. Sedo would appear ideal for installations intended for the long haul, where it would not be desirable to replace roof panels at 5- to 10-year intervals.

An alternative to Sedo is also available: double-glazed glass. The air space is usually 3.81 cm (1.5 in.), but since two pieces are used, algae build-up and condensation can occur. Based upon limited experience, fuel savings would appear to be comparable to those with double plastics (Ball 1981b).

Light transmittance through double-glazed glass is 80%. Longevity is nearly permanent, especially with a provision for flushing if algae should form. Mostly, double-glazed glass is similar in properties and problems to Sedo; its weight is between that of single glass and Sedo, as is its cost. At present only a few installations exist in the United States; most of these are renovations of existing greenhouses.

Future

The new roof materials discussed in this section need the test of time to determine which is the best. All save energy and as such will undoubtedly play some role in future greenhouse technology. New panels are in the research stage. These include glass combined with a Tedlar panel and plastics coated with additives to reflect infrared heat back into the greenhouse or to reduce condensates. There is even an experimental holographic coating designed to bend light to increase illumination in the dimmer areas (Aimone 1985). Improved greenhouse roof panels promise to hold the interest of researchers and growers for the immediate future.

Cost Payback Period

All the preceding options can save energy for greenhouse operations. The choice of which one to use will be determined by several factors: greenhouse condition, internal layout, geographic location, and greenhouse type. If a grower is lucky, evaluation of these factors will lead to at least two choices for the greenhouse in question. Growers can obtain evaluation help from the Cooperative Extension Service (see, e.g., Anon. 1982); in the near future, computer programs may be available for such evaluations (Bayles 1982). This brings the grower to the final determining factor, the cost payback period—that is, the time it will take to pay for the cost of the modification(s) based upon the savings generated from the improved energy efficiency.

The present unstable economy should make growers cautious about modifications that do not pay for themselves within about 6 years. Approximate cost payback periods for some of modifications were mentioned previously. However, due to the variability of conditions, individual calculations of the cost payback period should be made. A good explanation and some examples are provided by Botacchi (1980b). Major tax/depreciation savings are also applicable to greenhouse modifications that save energy (Ball 1982a, 1983).

Experimental Greenhouses

An interesting energy-efficient greenhouse is under study in Ohio (Short and Bauerle 1981). The greenhouse is covered with air-inflated double plastic rather than glass. At night polystyrene pellets are pumped between the two plastic layers. Their depth of 5 in. gives the structure an R value of 20, which is ten times the R value of DPOG! Such a structure would be extremely energy efficient and reduce fuel consumption by as much as two-thirds. A somewhat similar system is available commercially in Japan, but only for areas of little snow. Hopefully, such a system will become available in the near future in the United States as the researchers in Ohio resolve the snow load problem.

A greenhouse system combining solar energy for microclimatic heating and a thermal sheet is under study in New Jersey (Roberts and Mears 1981). The greenhouse covers 0.53 ha (1.3 acres) and consists of ten units connected gutter to gutter. The covering material is air-inflated double polethylene (4 mil). Five solar collectors heat water as it flows down the surface of a black plastic layer. Gravity flow feeds the water to the greenhouse floor, which is both a heat exchanger and water storage unit. The floor is essentially a vinyl-lined pool filled with gravel and capped with porous concrete. Water storage capacity is 378.8 kiloliters (100,000 gal.). Water is returned to the collector by a pump. Sensing devices determine the water flow direction. Auxiliary heat is available when needed. Crops are placed directly on the floor for direct root heating. A thermal sheet is drawn at night; the sheet is a woven polyester aluminized on one side. The porous nature of the sheet prevents water pooling from overhead dripping of condensate. A overhead heat line is used as needed to melt snow.

Energy consumption is 70% less in this experimental greenhouse than in a double-plastic greenhouse without solar collectors, a warm floor, and thermal sheet. The cost payback period is on the order of 6 years.

References

AIMONE, T.   1985.   Making the covering choice. Grower Talks *49* (1) 108–117.
ANON.   1982.   Greenhouse Energy Audit. Mass. Coop. Ext. Serv. Pamphlet.
BALL, V.   1985.   Fuel—some new answers. Grower Talks *49* (1) 68–76.
BALL, V.   1979.   Energy: a growing concern. Grower Talks *42* (12) 1–61.
BALL, V.   1981a.   Coal success story—John Van Bourgondien. Grower Talks *45* (5) 16–17.
BALL, V.   1981b.   New roof panels. Grower Talks *45* (5) 1–15.

BALL, V.   1982a.   Cut fuel bills without big investment. Grower Talks 46 (1) 14–43.

BALL, V.   1982b.   Temperature averaging: Trapping the sun's heat to save fuel. Grower Talks 45 (12) 1–9.

BALL, V.   1983.   Generating greenhouse heat. Conserving greenhouse heat. Grower Talks 46 (12) 20–41.

BARTOK, J. W.   1977.   Windbreaks reduce greenhouse heating costs. Conn. Greenhouse Newsl. 77 (January), 1–3.

BARTOK, J. W.   1980.   How efficiently is your heating system operating. Conn. Greenhouse Newsl. 97 (February), 36–38.

BARTOK, J. W.   1981.   Energy alternatives: Wood. Conn. Greenhouse Newsl. 104 (March), 1–3.

BAUERLE, W. T. and SHORT, T. H.   1977.   Conserving Heat in Glass Greenhouses with Surface-Mounted Air-Inflated Plastic. Ohio Agric. Res. and Devel. Center Circ. 101.

BAYLES, E. S.   1982.   Michigan State's energy-wise computer. Grower Talks 46 (5) 39–40.

BOTACCHI, A.   1980a.   Thermometer check. Conn. Greenhouse Newsl. 98 (March), 17.

BOTACCHI, A. C.   1980b.   Thermal blankets. Sealed glass laps vs. single poly over glass. Cost payback of systems for energy conservation. Conn. Greenhouse Newsl. 97 (February), 6–12, 28–30, 49–51.

GENT, M. P. N., THORNE, J. H. and AYLOR, D. E.   1979.   Split-night temperatures in a greenhouse: The effects on the physiology and growth of plants. Conn. Agric. Expt. Stn. Bull. 781.

HUGHES, J.   1981.   S.D.P., infrared heating, soil heating. Conn. Greenhouse Newsl. 104 (March), 9–13.

KOCH, P.   1981.   Ways to burn wood for industrial use. In Cutting Energy Costs (The 1980 Yearbook of Agriculture). J. Hayes (Editor). U.S. Dept. Agric., Washington, DC.

KOTHS, J. S. and SCHNEIDER, J.   1980.   Split-nite temperatures save energy. Conn. Greenhouse Newsl. 97 (February), 31–35.

KOTHS, J. S. and SCHNEIDER, J. S.   1981.   Split-night temperatures: Calceolaria. Conn. Greenhouse Newsl. 103 (January), 1.

LOEFSTEDT, W.   1977.   Comments on split-night temperatures. Conn. Greenhouse Newsl. 80 (July), 8–12.

MAISANO, J. J.   1980.   Latex paint on heating pipes. Conn. Greenhouse Newsl. 97 (February), 56.

PARUPS, E. V. and BUTLER, G.   1982.   Comparative growth of chrysanthemums at different night temperatures. J. Amer. Soc. Hort. Sci. 107 (4) 600–604.

REBUCK, S. M., WHITE, J. W. and ALDRICH, R. A.   1976.   Internal curtains reduce energy requirements in greenhouses. Penn. Flower Growers' Bull. 289, 1, 6–8.

ROBERTS, W. J. and MEARS, D. R.   1981.   Warm water solar system brings greenhouse saving. In Cutting Energy Costs (The 1980 Yearbook of Agriculture). J. Hayes (Editor). U.S. Dept. Agric., Washington, DC.

SALSEDO, C. A.   1980.   North wall insulation. Conn. Greenhouse Newsl. 97 (February), 24–26.

SCHNEIDER, J. S. and KOTHS, J. S.   1980.   Split-night temperature: Poinsettias, 1979. Conn. Greenhouse Newsl. 101 (September), 4–7.

SHORT, T. H. and BAUERLE, W. L.   1981.   Greenhouse production with lower fuel costs. In Cutting Energy Costs (The 1980 Yearbook of Agriculture). J. Hayes (Editor), U.S. Dept. Agric., Washington, DC.

STEFANIS, J. P., ALBRIGHT, L. D., WHITE, G. B. and LANGHANS, R. W.   1981.   A cost analysis of two energy conservation methods, Part II—Poly over glass. Conn. Greenhouse Newsl. 103 (January), 6–10.

THORNE, J. H. and JAYNES, R. A.   1977.   Split night-time greenhouse temperatures can save fuel. Conn. Greenhouse Newsl. 79 (May), 1–4.

WHITE, J.   1982.   Microclimate heating for energy conservation. Grower Talks 46 (1) 5–13.

Animal Husbandry Energy Conservation

Certainly the most criticized farm operation in terms of energy input versus output is livestock production. The efficiency of crop production is on the order of about 1% of the incident solar radiation; an additional 10-fold reduction of energy occurs when crops are fed to livestock. Critics feel it would be more energy efficient to feed crops directly to humans. Livestock producers and many meat consumers feel otherwise. An examination of the facts may help to put the issue into perspective.

First, agriculture consumes about 3% of our national energy; 90% of the energy used in agriculture goes for crop production and 10% for livestock operations. However, some crops are used for animal feeds. These include corn, soybeans, and alfalfa which rank first, second, and third in terms of acreage and first, second, and sixth in terms of energy consumption (see Table 3.1). Once this is factored in, livestock becomes the most energy-demanding product of U.S. agriculture (Table 5.1).

On the other hand, 80% of the energy fed to U.S. cattle is derived from forages unfit for human consumption, such as alfalfa, crop residues, non-grain plant parts, hay, and pasture. Of the grain that is used, not all is of high value or all that appealing to humans. High-moisture grain is fed directly (no drying) in cases where grain damage has occurred from drought, frost, mold, contamination, or kernel damage. Low-value byproducts from grain milling are also used as cattle feed (Fox 1981). Some of this would also be true for other livestock.

The counterpoint to these facts is that additional crops destined for human consumption could be grown on land utilized for forage and cattle production. However, much of this land is only marginal farmland. An additional point is that animal proteins have a greater nutritional value for humans than do single-plant protein sources. However, if plant proteins are combined in a selective manner and consumed, this argument loses ground. It must also be remembered that the efficiency for conversion of dietary energy to animal products varies; 10% is only an average value. Some livestock products have

Table 5.1. Average Energy Input for Animal Husbandry in the United States

Animal	Cultural energy (Mcal/1000 animals)	Percentage of total cultural energy				
		Feed water[a]	Fossil energy[b]	Machinery, equipment	Bldgs	Transportation
Cattle						
Dairy	5,866,000	65.8	26.7	6.5	—	—
Beef, range	899,017	33.8	—	37.5[c]	—	28.6[e]
Beef, fertilized pasture	629,208	92.0	2.9	3.87	—	—
Beef, feedlot (10,000 cap.)	2,077,291	85.9	13.1	1.0	—	—
Poultry						
Chicken, eggs	126,824	83.1	8.2	0.2	0.7	7.8
Chicken, pullets	27,894	77.8	6.5	0.4	8.8	6.3
Chicken, broilers	12,987	67.1	14.2	0.6	12.0	6.2
Turkey	146,058	75.8	11.9	0.1	5.8	6.3
Swine	650,500	64.8	25.1	10.1[d]	—	—
Sheep						
Range lambing	187,841	45.7	0.5	6.8[b]	—	47.0[d]
Shed lambing	426,442	22.0	2.5	10.6[b]	39.2	25.7[e]

Source: Adapted from data in Oltenacu and Allen (1980), Ostrander (1980), Reid *et al.* (1980), Cook *et al.* (1980), Hoveland (1980), Gee (1980).
[a]Includes production, processing, and distribution energy.
[b]Includes oil, gas electric, and some gasoline (but exclusive of gasoline used in transportation).
[c]Includes fence
[d]Includes buildings
[e]Includes pick-up and larger trucks

higher energy returns (Table 5.2); a change in the ratio of products consumed provides a viable alternative to elimination.

At present a powerful argument is that of reality. A demand exists in the market for livestock products. Even with increasing energy costs, a competitive, reasonably priced product exists. However as energy costs continue to rise, the incentive to reduce energy consumption in livestock operations becomes more powerful.

Environmental Maintenance

Control of the environment during livestock operations consumes approximately the same amount of energy as does heating of greenhouses: 0.039 vs 0.041 $\times$ 10^{12} MJ (CAST 1975; FEA 1976). Environmental control in livestock operations consumes energy for space heating, ventilation, and

Table 5.2. *Energy Efficiency of Converting Dietary Energy to Livestock Protein*

Product	Efficiency (%)	Product	Efficiency (%)
Eggs	10–14, 18	Chicken	11–16
Milk	10–20.5	Beef	2–5
Pork	5–14	Lamb	2

Source: Based upon data in Heichel (1976), Janick *et al.* (1976), CAST (1977), Cook *et al.* (1980), Gee (1980), Hoveland (1980), Oltenacu and Allen (1980), Ostrander (1980) and Reid *et al.* (1980).

brooding. As with greenhouses, a number of ways can be used to reduce energy input for environmental maintenance.

Production of hogs, some dairy cattle, and poultry is heavily dependent upon environmental control of confined areas, primarily to reduce labor and to improve production. Therefore, significant energy conservation possibilities exist in such operations, but caution is needed to not impair the environment to the extent that production is lowered. A good analysis of the problem has been presented by Stanislaw and Driggers (1981) and Thornberry (1981).

Insulation

Heating and cooling costs can be reduced through proper insulation of existing structures and new construction. Adequate ceiling or roof insulation (Fig. 5.1) reduces winter heating costs and minimizes the need for brooder fuel and feed. During the summer it helps to reduce the need for ventilation and cooling, and to minimize problems of temperature stress. In cold climates wall insulation can also be profitable. The cost payback period for maximal insulation is 5–7 years or less depending on tax breaks, making it a worthwhile investment for hog and chicken production, and for dairy cattle kept in buildings.

The goal of using insulation, weatherstripping, and polyethylene film over windows is to reduce heat loss to a maximum of 5 Btu/hr/ft^2. The required R value of the insulation would be 8 to 16, depending upon the expected difference between inside and outside temperatures. This is equivalent to about 2–4 in. of mineral wool, 1.5–3 in. of foam-type insulation, or 3–6.5 in. of impregnated fiber board. The difference in energy consumption between a well-insulated hog or poultry house (heat loss of 5 Btu/hr/ft^2) and a poorly insulated house (heat loss of 15 Btu/hr/ft^2) can be as much as threefold.

Livestock operations, especially dairy operations, often require hot water for cleaning equipment and sanitation. Insulation of the hot water heater can reduce energy costs by up to 10%. Hot water lines can also be insulated and should be if they pass through unheated areas.

Waste Heat Recovery

The cooling of milk in dairy operations requires a bulk tank compressor. This unit emits substantial amounts of low-temperature heat, half of which can be recovered with commercially available heat reclaimers. The reclaimed heat can be utilized to heat water for cleaning purposes, thus cutting fuel needs by 40–50%. Since water heating requires roughly 20% of the energy consumed in dairy operations (FEA 1976), an overall energy decrease of 8–10% is possible.

Cost payback periods are only a few years, making this option quite attractive. Certainly waste heat recovery and insulation of the hot water heater and lines should be attempted before considering solar hot water heaters.

FIG. 5.1. Ceiling insulation in poultry houses can reduce radiant heating by the sun, thus reducing energy costs associated with ventilation and cooling. In the winter this insulation slows heat loss, thereby decreasing fuel requirements.
Courtesy U.S. Department of Agriculture

Ventilation

In some instances it may be possible to replace mechanical ventilation with natural ventilation. Such is the case with open sidewall poultry housing or buildings used for partial periods of confinement, as in dairy operations. During the winter mechanical ventilation may still be required for better temperature control, but savings would be possible in the warmer months if the structure were opened and naturally ventilated. When mechanical ventilation is used, rates should be minimal, but safe, during the heating season. Fans should also be turned off when ventilation is not required.

Rates of ventilation are especially important. Excessive ventilation in the winter increases heating costs, since the incoming cold air requires warming. Too little ventilation leads to reduced quality of hogs and chickens. Another important consideration is the interaction between heating, ventilation, and moisture-holding capacity of the air.

Animals, wastes, and litter generate moisture, which must be controlled. Normally, excess moisture is removed through ventilation. However, the moisture-carrying capacity of air is a function of its temperature; it decreases roughly one-half with a drop of 11°C (20°F). Therefore, as the temperature drops, the volume of incoming air must be raised, and even though it must now be heated to a lower temperature, there is a greater amount to bring up to the interior temperature. Depending on interior and exterior conditions and the set relative humidity, a temperature drop results in little to no decrease in fuel consumption; it may even increase energy consumption. As an example, a drop in temperature from 21° to 10°C (70° to 50°F), with a set relative himidity of 75%, requires increased ventilation and a 40% increase in heating (Stanislaw and Driggers 1981).

For the best energy efficiency and optimal hog productivity, the ideal temperature is 21°–27°C (70°–80°F) with ventilation rates designed for good air distribution and proper moisture removal. This becomes especially important during the heating season in enclosed hog and chicken buildings.

Other factors can also be helpful in minimizing the energy requirements of ventilation. Fans should be chosen on the basis of efficiency; fans may differ by as much as 6.6 cfm/watt. The wrong choice could increase annual energy cost by a hundred dollars or so per fan (Thornberry 1981). Large-diameter, slow-speed fans with lower horsepower are usually 5–15% more energy efficient than high-speed fans (greater horsepower) of similar air movement capacity. Dust on fan blades or ventilation louvers can cut air flow and energy efficiency by up to 10%, so they should be cleaned periodically to maintain maximum energy efficiency. Fans should have enclosed motors to prevent damage from dust and humidity. Because hog and chicken operations generate considerable dust, frequent inspections and cleaning are required.

Evaporative Cooling

In warmer parts of the United States, evaporative cooling is required along with ventilation, especially in poultry operations. Effective roof or ceiling insulation is a must for energy efficiency with evaporative coolers. Pads must be kept moist at all times and replaced promptly upon deterioration for energy-efficient operation. Periodic cleaning of nozzles and replacement of filters are helpful in maintaining constant wetness. Water costs can be decreased by recirculation of cooling pad water and by a constant watch for leaks and runover. Dust and debris should be routinely removed such that air flow through wet pads is not reduced.

Lighting

Some modifications of light systems and schedules can conserve energy. For example, constant high-intensity light is not needed for market poultry. By adoption of a intermittent schedule of low-intensity light (9.5 foot candle at bird level), electrical energy usage can be cut (Thornberry 1981). Changes in duration or intensity of lighting, however, should be initiated with new flocks and not those already into egg production. Fluorescent lights are more efficient than incandescent bulbs and might be considered for new construction.

Other simple steps can also help reduce electrical costs. Placement of aluminum reflectors and periodic cleaning of bulbs can cut electrical energy use by 25%. One 25-watt incandescent light with a reflector is the equivalent of a clean 40-watt bulb without reflector or a dirty 60-watt bulb (Thornberry 1981). Other steps include the use of spot lighting of heavily used areas, as opposed to whole area lighting. Timers or photo-cells can also be used to insure that lights are only on when needed.

Brooding and Farrowing Areas

Two operations requiring heavy energy expenditure for environmental control are the farrowing of swine and the brooding of poultry. The largest part of the energy for environmental maintenance with poultry is used for brooding operations (Thornberry 1981). Energy costs for production of feeder pigs (farrowing and nursery operations) are at least two and one-half times greater than those for the later phase of feeding out (Stanislaw and Driggers 1981). Thus these operations are prime candidates for energy reduction.

The costs for hog farrowing and poultry brooding can be reduced through proper insulation and ventilation, as discussed already. In addition, a number of energy-efficient brooding practices can be utilized with poultry. One of the most promising in terms of energy reduction is partial brooding (Fig. 5.2), which can reduce energy requirements 25–66% (Thornberry 1981). With this practice chicks are restricted to 15–30% of the house by a floor to ceiling plastic curtain, which is left in place 10–14 days; then the curtain is moved to divide the house in half for another 10–14 days. During this time no heat is required for the unused portion, except to prevent freezing of pipes.

In the much more open poultry housing common to the South (curtain sidewall as opposed to enclosed in North), fuel use during brooding can be cut by 10–15%. A layer of polyethylene is attached to the inside of the sidewall opening to reduce air leakage. If ventilation is natural by opening of the sidewall curtain, the plastic should not extend completely to the top of the sidewall.

Adjustment of temperature in brooding areas will also conserve energy. Temperatures below the usual brooding temperature of 32°C (90°F) have been used with success: 29° to 30°C (84° to 86°F). Strict care must be used to avoid temperature fluctuations and excessive humidity and moisture. If high humidity is a problem, the usual temperature of 32°C is better. Stress can also necessitate the need for a higher temperature. An alteration of temperature reduction is also helpful for saving fuel. Instead of a drop of 3°C

FIG. 5.2.   The use of a plastic curtain to partition a poultry house for partial brooding is a good way to reduce fuel consumption. Only the smaller area needs to be heated to the higher brooding temperature and poses no crowding hazard to the poults.
Courtesy U.S. Department of Agriculture

(5°F) each week, it appears that a drop of 1° to 2°C (2 to 3°F) every 3 to 4 days is acceptable (Thornberry 1981).

Other minor steps can also help conserve energy. Brooders should be maintained on a scheduled basis for optimal operation. Pilots should be turned off when brooders are not in use. Brooders should be used to their fullest, not partial, capacity. Thermostat accuracy should be verified with a reliable thermometer to avoid fuel waste. Finally, when new brooders are purchased, both fuel-efficiency ratings and cost should be considered.

Feed

The production of feed is costly in terms of energy. For example, 86% of the energy consumed in beef production is used for feed production and 14% for feedlot operation. There would appear to be a compelling case for energy reduction during feed production. The elimination of grain would appear to be the simplest answer, but as stated previously 80% of all cattle feed energy is derived from rangeland forages, a renewable resource not readily suitable for human consumption. However, a number of steps can be taken to reduce energy input into cattle feed (Fox 1981), besides the limited decrease possible with forages.

Cattle Feed

About 151.5 liters (40 gal.) of fuel are required to produce 0.9 metric ton of cattle feed. Growth stimulants, used properly and in optimal amounts, can save an energy equivalent of 151.5 liters of fuel for four steers fed from weaning to normal slaughter weight. This savings can be doubled to 75.8 liters (20 gal.) of fuel per steer if feed additives that improve feed efficiency are used in conjunction with growth stimulants. Of course, this approach only works if the diet is carefully balanced to avoid nutrient limitations or excesses. Poor ration formulation in itself can increase feed use by as much as 10%.

Another alternative is to supplement cattle diets with byproduct feeds. A number of byproducts with feed value are produced by certain food processors (cereal, fruit, vegetables), brewers, and distillers. These byproducts include wheat middlings, hominy, corn gluten, molasses, cottonseed meal, wheat mill run, and brewers/distillers grain. Greater use of these byproducts could further reduce energy costs. Farm alcohol production also yields distillers grain.

An example of their effectiveness is shown by a comparison of dairy cattle raised in the Southeast and Lake States. Lake State dairy operators use an average of 6% byproducts in their feed, whereas those in the Southeast use

31%. Consequently, only 71% as much feed energy is used for dairy cattle by the latter. In all fairness, part of this difference is also attributable to differences in pasture usage: 1% in the Lake States and 6% in the Southeast.

Grains should be handled with a view toward energy conservation. High-moisture grain can be stored and fed directly instead of being dried. Long-term storage of grain requires some form of preservation, such as fermentation or preservation with propionic acid sprays. Use of high-moisture grain requires more careful management of storage and some alterations in diet formulation, but yields a net savings of energy. Losses normally incurred during harvesting are reduced, since grain not to be dried can be harvested earlier. About 0.7 liter of LP gas is saved for each bushel of corn not dried. The energy costs of grain processing should also be considered. For example, steam flaking of corn increases feeding efficiency and results in less required feed, but the processing energy exceeds that of the energy required to produce the additional feed needed without processing. Processing is energy efficient at other times: with low-moisture corn (less than 10% moisture), diets with roughage exceeding 50%, cattle older than 10–15 months, and with small grains.

Steps to conserve energy are possible for cattle in the pasture. For example, grass pasture plantings can be altered to include legume forages. Besides being good protein sources, less nitrogen fertilizer will be needed. Fuel costs for fertilizer (production, transport, and application) are substantial; each kilogram of fertilizer not used saves the equivalent of 2 liters of gasoline. The use of legumes in pastures could cut fertilizer requirements by 25–40%. Grazing of nonirrigated, fertilized pastures, especially pastures incorporating legumes, is at least eightfold more energy efficient than feeding harvested feeds.

Rotational grazing, in which cattle are confined to a series of smaller areas, requires less time and fuel for supplemental feed distribution and cattle checks than does continuous grazing. Further fuel savings are possible by alternate rather than daily distribution of supplemental feed. Rotational grazing also usually improves pasture production and carrying capacity. The rotational scheme and less frequent distribution of supplement could save 4000–5000 liters (15,200–19,000 gal.) of gasoline annually on large ranches with widely scattered cattle.

Feed for Other Livestock

The energy associated with feeding hogs, poultry, and lambs is less than that for cattle. Nonetheless, there are some simple ways to conserve energy in the feeding of livestock other than cattle. In some cases the energy cost for pelletization of feeds for poultry and older lambs may cost more than the energy realized from the increase in nutritional efficiency. Increased feed storage may be less costly than energy costs associated with more frequent feed deliveries. Alternate feeds, such as corn silage, could be substituted for more energy costly pelleted feeds used with lambs (Thornberry 1981; Wickersham 1981).

Feed Processing and Handling

Some energy savings are also possible in the processing and handling of dairy and beef cattle feed (Stewart and Davis 1981). Feed grinding and mixing consumes less energy if lower horsepower machinery is used over a longer period of time. Assuming the operation is large enough, three-phase electrical service can improve system efficiency, and three-phase motors cost less and are more energy efficient by 5% than single-phase motors. Wherever possible, feed should be distributed with conveyors and augers, rather than vehicles. Self-feeding and gravity flow, rather than mechanized systems, should be used when feasible. Finally, equipment should be properly maintained for optimal efficiency.

Transportation

Energy costs for travel, assembling, and handling by sheep producers are proportionately greater than the same costs associated with other livestock raised for meat (Table 5.1). One form of energy conservation available to sheep farmers is use of the working sheep dog. Such trained dogs can gather and return a sheep flock for distances of around 0.5 km (0.3 mile). This cuts down on gasoline or diesel use in pickup trucks or small tractors normally used to gather sheep (Wickersham 1981).

The gathering of sheep can also be cut from several to three or four times per year through more efficient management, thus cutting energy costs. This improved efficiency can be achieved by carrying out several operations, rather than just one, during each gathering. Common operations include annual shearing, worming twice or more per year, feet trims, sorting and marketing, and vaccinating.

Energy costs for assembling and handling livestock have escalated as the number of slaughter plants has declined and their distance from many producers increased. Sheep producers have reduced their transportation costs by pooling. Lambs are assembled from various producers at a collection point at most 50 km (30 miles) away from each and then transported to the slaughter house in large, fully loaded trucks. Similar procedures can be used when replacement ewes are needed by several producers (Wickersham 1981).

Other attempts at energy conservation can reduce transportation expenditures in all livestock operations. Engines should be properly tuned, lubrication schedules followed, tires inflated to correct pressures, and good driving habits encouraged. Avoid rapid starts, use minimal braking, accelerate smoothly, maintain moderate speeds, and do not idle the engine during prolonged stops. All these steps will help motorized vehicles to achieve and maintain optimal fuel economy. Unnecessary trips should be avoided and two or more potential trips combined whenever possible.

New vehicles should be purchased based on two requirements: high fuel economy and a good match of vehicle size and horsepower to intended purposes. If possible, manual transmissions should be sought, since they have a slight edge over automatic transmissions. Options that cut fuel economy, such as air conditioning, should not be purchased unless deemed necessary.

Waste Disposal

Often wastes from livestock can be used to save energy by being substituted for chemical fertilizers for on-site crop production. Excrement from a steer over a 240-day feeding period produces about 33 kg (72 lb) of nitrogen, which is equivalent to 60 liters (15.8 gal.) of gasoline (Fox 1981). As discussed in Chapter 1, substitution of manure for nitrogen fertilizer can result in significant energy savings if it is used directly on the farm site of origin. Most manure is already used as fertilizer on pastures or for crop production, especially on mixed crop—livestock farms. However, hauling costs from feedlots to farms can cancel out savings, leaving feedlot manure unused. Feedlot manures present a waste disposal problem; it would make good sense to both dispose of and recycle plant nutrients at the same time.

Fortunately, manures can be used as feedstocks to produce energy and nutrients. Every kilogram (2.2 lb) of manure yields 0.44 cubic meters (15.4 ft³) of gas through anaerobic digestion. This gas contains 60% methane and has a heat content of 17,650 Btu/m³ (500 Btu/ft³). The residue left after anaerobic digestion can be slurried with water and then fermented in a digester at 93.3°C (200°F) with suitable bacteria to yield more methane. The heat needed for digestion can be supplied by a fraction of the generated methane. The protein-rich residue left makes an excellent fertilizer or, if dried, possibly an animal feed ingredient (Miller 1981).

Such a system is feasible and may become common if energy costs continue to climb. Estimates of manure's value for direct conversion to energy are very reasonable. One hundred dairy cows with an average weight of 681.8 kg (1500 lb) produce enough manure to generate a daily output of 1.6 million Btu, equivalent to 45.5 liters (12 gal.) of diesel fuel. This estimate assumes a 60% system efficiency for conversion of manure to methane (Stewart and Davis 1981). Methane produced at a large feedlot could be sold to a gas utility and the resulting sludge sold or used as a fertilizer.

The economic value of this sludge would be based on its nitrogen content; nitrogen is currently priced at about 32¢/lb (70¢/kg). Sludge would also probably be more efficient than regular manure because the nutrients are concentrated and in a more easily handled form. About half the nutrients are lost through handling of manures before they reach the soil. The sludge form should increase nitrogen conservation.

Small-scale methane production could be used to heat farrowing houses for swine or brooding houses for poultry, or to provide hot water for cleaning needs. For example, sewage treatment plants have produced methane from anaerobic digestion of human wastes and used it for heating purposes for many years. In this way natural gas, a nonrenewable resource, could be replaced with a renewable gas, methane, thus helping to sustain agriculture. Whether methane production and use on farms becomes common will depend upon several factors: conventional fuel costs, economy of scale, and value of residues for animal feed and fertilizer.

Disposal of livestock wastes incurs an energy expenditure, but some forms of disposal are more energy intensive than others. For example, in livestock confinement quarters, a daily cleanout with a mechanical scraper system offers distinct advantages in terms of odor and insect control but is energy intensive in terms of maintenance and operation. On the other hand, an in-house system of allowing the litter to build up and compost not only is less energy intensive in terms of cleanout, but the heat from the biological oxidation may help to offset heating needs, such as during poultry brooding (Thornberry 1981). Other forms of waste disposal may also be less energy intensive than daily cleanout, but some (e.g., lagoon-flush systems) may result in loss of fertilizer value. Still others (e.g., in-house dry storage) require the use of insecticides for insect control and increased ventilation, which in turn increase energy consumption.

Regardless of what system is chosen, energy conservation depends upon proper and regular scheduling for maintenance of disposal equipment. A second element is that the system of choice should be one that conserves plant nutrients, so that further energy savings are possible. If livestock wastes are not used for fertilizer, their conversion to methane may be warranted (see Chapter 10).

References

CAST. 1975. Potential for Energy Conservation in Agricultural Production. Rept. No. 40. Council for Agric. Sci. and Technol., Dept. of Agronomy, Iowa State Univ., Ames.

CAST. 1977. Energy Use in Agriculture: Now and for the Future. Rept. No. 68. Council for Agric. Sci. and Technol., Dept. of Agronomy, Iowa State Univ., Ames.

COOK, C. W., COMBS, J. J. and WARD, G. M. 1980. Cultural energy in U.S. beef production. *In* Handbook of Energy Utilization in Agricultural Production. D. Pimentel (Editor). CRC Press, Boca Raton, FL.

FEA. 1976. Energy and U.S. Agriculture: 1974 Data Base. FEA/D-76/459. Fed. Energy Admin.–U.S. Dept. of Agric., Washington, DC.

FOX, D. G. 1981. How to produce beef for less money. *In* Cutting Energy Costs (The 1980 Yearbook of Agriculture). J. Hayes (Editor). U.S. Dept. Agric., Washington, DC.

GEE, C. K. 1980. Cultural energy in sheep production. *In* Handbook of Energy Utilization in Agricultural Production. D. Pimentel (Editor). CRC Press, Boca Raton, FL.

HEICHEL, G. H. 1976. Agricultural production and energy resources. Amer. Sci. *64* (1) 64–72.

HOVELAND, C. S. 1980. Energy inputs for beef cattle production on pasture. *In* Handbook of Energy Utilization in Agricultural Production. D. Pimentel (Editor). CRC Press, Boca Raton, FL.

JANICK, J., NOLLER, C. H. and RHYHERS, C. L. 1976. The cycles of plant and animal nutrition. Sci. Amer. *235* (3) 75–86.

MILLER, D. L. 1981. The ABC's of making farm alcohol and gas. *In* Cutting Energy Costs (The 1980 Yearbook of Agriculture). J. Hayes (Editor). U.S. Dept. Agric., Washington, DC.

OLTENACU, P. A. and ALLEN, M. S. 1980. Resource-cultural energy requirements of the dairy production system. *In* Handbook of Energy Utilization in Agricultural Production. D. Pimentel (Editor). CRC Press, Boca Raton, FL.

OSTRANDER, C. E. 1980. Energy use in agriculture poultry. *In* Handbook of Energy Utilization in Agricultural Production. D. Pimentel (Editor). CRC Press, Boca Raton, FL.

REID, J. T., OLTENACU, P. A., ALLEN, M. S. and WHITE, O. D. 1980. Cultural energy, land, and labor requirements of swine production systems in the U.S. *In* Handbook of Energy Utilization in Agricultural Production. D. Pimentel (Editor). CRC Press, Boca Raton, FL.

STANISLAW, C. and DRIGGERS, B. 1981. How to raise hogs for less money. *In* Cutting Energy Costs (The 1980 Yearbook of Agriculture). J. Hayes (Editor). U.S. Dept. Agric., Washington, DC.

STEWART, L. E. and DAVIS, R. F. 1981. An energy-saving list for dairy production. *In* Cutting Energy Costs (The 1980 Yearbook of Agriculture). J. Hayes (Editor). U.S. Dept. Agric., Washington, DC.

THORNBERRY, F. D. 1981. Some better ways to raise poultry. *In* Cutting Energy Costs (The 1980 Yearbook of Agriculture). J. Hayes (Editor). U.S. Dept. Agric., Washington, DC.

WICKERSHAM, T. 1981. How to raise sheep easier and cheaper. *In* Cutting Energy Costs (The 1980 Yearbook of Agriculture). J. Hayes (Editor). U.S. Dept. Agric., Washington, DC.

6

Postproduction
Energy Conservation

We have already examined agricultural energy input for crop and livestock production. These segments of agriculture together account for 18% of the energy utilized in the food system. The remaining energy usage, 82%, occurs beyond the farm gate. Postproduction energy is needed for transportation, processing, marketing, and preparation of food. At least 75% of U.S. crops and livestock are processed in some way before marketing to consumers.

Postproduction energy usage in the food system presently accounts for around 13.5% of the total U.S. energy consumption and ranks sixth in gross energy use for U.S. industry groups. A breakdown of the amount and cost of energy used in the postproduction phase of the U.S. food system is shown in Table 6.1.

Energy costs are passed on to the consumer who pays, on the average, about ten cents for energy out of every dollar spent on food. Some foods have higher energy costs and others lower, as indicated in Table 6.2. Much of this energy is used in processing activities, which are highly dependent upon oil, natural gas, electricity, and coal.

A few factors have contributed to the increasing postproduction energy

Table 6.1. Annual Postproduction Energy Use and Cost in the U.S. Food System

Activity	Energy use		Cost of energy	
	(Btu $\times$ 10^{15})	(%)	($ $\times$ 10^{9})	(%)
Food processing	3.6	35.3	8.1	25.9
Marketing and distribution	1.3	12.7	4.6	14.7
Wholesale		3.7		
Retail		6.0		
Commercial eating establishments	2.1	20.6	7.1	22.7
In-home preparation/use of food	3.2	31.4	11.5	36.7

Source: Based upon USDA (1979) and Vilstrup (1981).

Table 6.2.   Unit Energy Cost and Annual Total Energy Used for Processing Selected Food Groups

Food groups	Energy cost (cents/dollar of processed product)	Energy use (billion MJ)
Meat products	9.7	823
Fluid milk	8.6	330
Bakery products	6.7	294
Alcoholic beverages	6.3	251
Canned fruits and vegetables	10.6	178
Butter, cheese, condensed milk	10.7	139
Soft drinks	8.5	118
Frozen fruits and vegetables	10.5	99
Flour and cereals	9.9	83
Ice cream	8.7	73
Sugar	16.7	71

Source: Based upon Hirst (1973) and Vilstrup (1981).

usage of the food system. Demand for processed food in the most convenient form and meals outside the home have increased as the majority of families have both fathers and mothers working outside the home. Export demands for food are also high, as the United States continues to lead in agricultural exports. Much of the energy here is needed for transportation, since most of agricultural exports are not processed to any great extent. Socioeconomic conditions are such that it is unlikely that energy conservation can come about through reduction or elimination of these factors. Consequently, the choice is more efficient use of energy. This approach requires a knowledge of energy patterns in various postproduction activities.

Food Processing

Energy needs of food processors are critical at certain times of the year. Timely operations are of prime importance if the quality of perishable agricultural products is to be maintained. Timeliness is extremely critical in the canning and freezing of fruits and vegetables. These processing plants operate at maximal output for minimal periods of time (6–12 weeks); energy shortages at such a time would have devastating economic consequences.

Fuel stockpiles are not the answer for many food processors. Located in urban areas with premium land values, zoning restrictions, and consideration of neighborhood resistance, many plants cannot store oil or coal in bulk. Instead they are dependent upon gas pipelines. Of fourteen leading food processing industries, thirteen were dependent upon natural gas for at least one-third of their energy requirements (Unger 1975). The amounts of various fuels used in food processing are shown in Table 6.3. Total energy use in food processing accounts for 7.6% of all manufacturing energy.

Energy consumption and energy intensity vary considerably among products. For example, processing of meat products (Fig. 6.1) is the number one energy user in terms of total consumption (Table 6.2), but not the most energy-intensive food industry. Beet sugar processing is the most energy-intensive industry, followed by wet-corn milling (Singh 1981). This energy intensity is reflected in unit costs; for example, the cost attributed to energy in each dollar spent on product is nearly twice as high for sugar than it is for meat.

Few studies have been conducted on the comparative energy consumption of different forms of preservation used by food processors. Brown and Batty (1976) reported that the manufacture of a can for corn consumed 1.4 times as much energy as the production of corn. The energy cost of a can may be as high as 1.9 times the energy used to process the contents (Singh 1981). Thus, the processing and packaging (can plus carton) of corn requires four times as much energy as is required for corn production (Brown and Batty 1976).

Freezing, an alternative form of preservation, has a variable energy cost, dependent upon the time frozen products are held until consumption. For example, if frozen corn is held roughly 3 weeks, its energy cost exceeds that of canned corn; if held 1 year, the energy cost becomes 17 times as great (Fluck and Baird 1980). Another study by Rao (1980) showed that energy consumption of a 2.9-oz serving of corn kernel was 2937, 2541, and 2875 Btu for fresh refrigerated, frozen, and canned corn, respectively. This included energy consumption after harvest to before in-home preparation, but essentially no home storage. Thus, frozen foods can consume large amounts of energy for storage in wholesale, retail and home; on average, 3.4 times as much energy is used to store frozen foods than to process them (Singh 1981).

Table 6.3. *Annual Use of Various Energy Sources in the Food Processing Industry*

		Energy equivalent	
Source	Quantity	(Btu $\times$ 10^{12})	(%)
Oil, distillate	11.5 $\times$ 10^6 barrels	67.0	9.2
Oil, residual	13.9 $\times$ 10^6 barrels	87.4	12.0
Coal	3.9 $\times$ 10^6 short tons	109.2	14.9
Natural gas	441 $\times$ 10^9 cubic feet	466.1	63.9
Electricity	38.3 $\times$ 10^{12} kWh	0.011	0.002

Source: Adapted from Vilstrup (1981).

FIG. 6.1. The largest energy consumer in the U.S. food industry is the meatpacking industry, which consumes more than 99 trillion Btu each year.
Courtesy U S. Department of Agriculture

The energy efficiency of other food preservation methods has not been examined closely. These other methods include dehydration, pickling, fermentation, concentrating (jelly and jams), radiation sterilization, and chemical additives. Some limited information is available on dehydration. Production of dehydrated fruits and vegetables requires on the average 25% more energy than frozen forms and 80% more than canned forms (Pierotti *et al.* 1977). The increase in energy usage associated with various methods of preserving potatoes, using the fresh form as the base, is as follows: flake-dried dehydration, 9.4%; microwave dehydration, 43.5%; canning, 72.8%; freeze dried, 147.6%; and freezing, 190% (Olabode *et al.* 1977).

Some forms of food preservation are more energy competitive than others. Additional studies are needed to give a complete assessment; the most energy costly steps need careful evaluation if reliable attempts to improve energy efficiency are to be forthcoming.

Alternate forms of energy for food processing are being examined. Only limited use of solar energy occurs at present, since cost payback periods are unattractive. Solar collectors are presently too expensive and have poor efficiencies. The most efficient use of solar energy at present is for drying of onions, garlic, grapes, and apricots, and for heating of water for processing and cleaning purposes. Some plants are using their waste products, such as shells and pits, to fuel steam boilers. One canning company estimates that the burning of peach pits to produce steam will cut annual fuel costs by $190,000 (Singh 1981).

At present it appears that conventional forms of energy will continue to be used in the food processing industry. Nonetheless, adoption of various conservation measures resulted in a 17% increase in energy efficiency in the food industry from 1972 to 1978. These conservation measures included reduction of excessive lighting, improvement of boiler efficiency, and installation of steam traps and automatic heating controls.

Energy Accounting Techniques

The first requirement for energy conservation in food processing is an energy accounting technique to provide quantitative information on energy use throughout the food processing system and determine the relative energy input for each step. Such information would provide a base for energy conservation approaches. An energy accounting technique of this type was formulated by Singh (1978). It consists of a series of steps and the use of energy sensors to monitor energy use, followed by careful data analysis. A more recent model for assessing energy conservation potential during food processing is that of Levis *et al.* (1981).

Griffith *et al.* (1979) used an energy accounting technique to examine an atmospheric retort that is required for sterilization of canned foods. The resulting modifications, primarily in the heat exchangers, suggested by this analysis produced a 50% reduction in energy use and annual savings of $59,000, with a cost payback period of one season. These modifications are being implemented by many canners now. Singh *et al.* (1980) conducted an energy accounting of the canning of whole, peeled tomatoes. They pinpointed several parts of the process that consumed extensive amounts of energy and would be prime candidates for energy conservation.

The one possible drawback with the energy accounting procedure is that costs can be expensive for large-scale audits. However, savings may be substantial enough to result in an economically feasible cost payback period. As energy costs continue to climb, energy accounting techniques will undoubtedly see more widespread use.

Equipment

Most of the food processing equipment currently available was designed in the times of cheap energy, so there is a potential for design improvements to bring about energy savings. Hopefully, industrial and university research will result in much more energy-efficient food processing machinery, which will gradually replace existing equipment over the long term. Engineering redesign already has led to more energy-efficient equipment for cold and freezer storage. This type of storage is an important part of food processing and distribution and is covered well by Hallowell (1980).

In the short span it may be possible to modify existing equipment and associated operating procedures (Inaba et al. 1981). Energy accounting methods will be helpful for determining the most promising directions for both modification and redesigning of food processing equipment. At present very little reliable data exist on energy consumption by such equipment.

Miscellaneous Energy Conservation Measures

Food processing at commercial sites has been examined closely in terms of energy conservation possibilities over the last few years, resulting in a number of improvements. Fruit processing often utilizes a dehydration step, which is costly in energy. Tunnel dehydrators are used to dry fruits, primarily to produce raisins, prunes, and dried apples. Thompson et al. (1981) showed that certain energy conservation techniques (reduction of air leakage, increased air recirculation, use of a furnace heat shield, and maximization of fruit loads) could produce significant energy savings. Potential maximal energy savings for each technique, respectively, were 3–4, 15, 10, and 25%.

Citrus processing plants consume the most energy during the removal of water from juice and peel. Usually the juice is concentrated in steam-heated evaporators and peels are dried in rotary driers followed by waste-heat evaporators. Leo (1982) showed that conversion to cogeneration was economically feasible and would produce considerable energy savings. Cogeneration is the simultaneous generation of work (usually in the form of electricity) and process heat or steam from the same fuel. Systems capable of cogeneration examined by Leo included the gas turbine, the steam turbine, and the reciprocating engine. Energy savings with these were 29, 10, and 12%, respectively, compared with a standard boiler system. The gas turbine was best, as it had a cost payback period of 3 years and seemed adaptable to citrus processing. A more immediate energy savings of 3–10% could be achieved through the use of microcomputers to automate existing evaporators.

Soybean processing for purposes of making vegetable oil and livestock feed has been improved. A new, one-vessel extraction system (Morris 1979) increased production capacity by 20% at an Indiana soybean oil plant and cut energy needs by 15–20%. Processing capacity was 2400 tons of soybeans per day. In this system soybeans are desolventized, toasted, dried, and cooled to ambient temperature. The entire operation is automated through the use of computers. Besides being energy conservative, the system is also environmentally better than previous extraction systems.

Blanching of vegetables prior to freezing is more energy efficient if the water is recycled. The recycling of water also conserves water and reduces pollution; the latter results fom a reduction in solids loss. With a recycling process described by Swartz and Carroad (1979) vegetable quality remained good if the pH of the recycled water was kept near neutrality. A prototype for a new blanching system promises a 20-fold reduction in energy requirements according to Cumming (1980). The process is called the K-1 system and utilizes an individual quick-blanch approach. Leaching is minimal and ascorbic acid retention is improved. Future use depends upon improvement of handling capacity, but commercialization is feasible.

Mann (1981) has reviewed developments on energy conservation in the processing of dairy products. A large number of approaches were covered, many of which could bring about considerable reductions in energy consumption and had reasonable cost payback periods. These steps varied from modifications of existing processes to new techniques. Energy savings varied, with some being as high as 95%. Some examples are as follows.

Among the modifications in milk processing cited by Mann (1981) are practical means to cut heat requirements. For example, friction heat at the diversion value could raise the milk temperature nearly 3°C (5°F). Such heat could be utilized by repositioning the homogenizer before the flow diversion value. The flash cooling of cheesemilk during pasteurization generates waste heat, which could be reclaimed and used to preheat the raw milk used to make cheesemilk. Heat reclaimers were also suggested for recovering waste heat from refrigerator condensors.

If the existing refrigeration system is old and energy inefficient, the installation of a new system may result in considerable savings. Mann (1981) cites the example of a St. Louis dairy plant, where a new refrigeration system produced a $24,000 annual reduction in energy costs. An additional savings of $180,000 resulted because of decreased costs for labor, maintenance, and other related operating costs.

The use of cogeneration also offers considerable energy reductions during the processing of raw milk into pasteurized milk, butter, and cheese. A German dairy processor (Mann 1981) installed a cogenerator system and cut energy costs by 22%. The cost payback period was about 3½ years.

Energy costs during homogenization can be cut substantially by a change in the processing method. An older, but unused process is available for separating whole milk into light cream and skim milk. It is possible to homogenize only the light cream fraction and then recombine it with the skim milk to produce homogenized whole milk. Since the light cream fraction is only 20% of the volume of whole milk, smaller homogenizers can be used.

Valentin (1980) has covered the improvement of energy efficiency in the beet sugar industry. The use of gas turbines with waste heat boilers offers the most energy-efficient processing system. Existing systems could reduce energy costs by utilizing waste heat from the vacuum pans and the pulp driers. Valentine recommends heat pumps for this purpose. Partial replacement of the thermal drying process by mechanical dewatering to 30% solids requires about one-third less energy.

Energy conservation in refrigerated storage systems has also received considerable attention (International Institute of Refrigeration 1978; Forwalter 1979; Rutlin 1980). One area of interest is the recovery of waste heat from refrigeration uints in food processing; this heat can be used to provide hot water at 60°C (140°F). The cost payback period is 2 years, making this an attractive energy conservation measure.

Food Waste Recovery

Food processors generate waste products which represent lost energy and a source of pollution. Therefore another direction toward energy conservation would be to devise more efficient means to utilize materials during food processing. This approach has been reviewed by Moon (1980) and Knorr (1983).

Actually wastage is incurred throughout the entire journey of food from the farmer to the consumer. Losses occur during cultivation, harvesting, storing, processing, marketing, preparation, cooking, and consumption. For example, Knorr (1983) points out that the total wastage for potatoes from cultivation through consumption is 15.2%. However, we will be concerned primarily with the wastage during processing. The reason for this focus is that food is more concentrated at the processing end; thus it is more economical and feasible to apply waste abatement procedures.

Some generalizations are possible from an examination of wastage data derived from food processing (Moon 1980; Knorr 1983). For example, the canning of fruits results in an average waste of roughly 9% of the edible material; if wastage is calculated on the basis of all material, edible and inedible, the value is about 43%. The respective figures for vegetables are quite similar, 9 and 45%.

The lowest losses of edible material occur with citrus fruits (3%) and the highest with peaches and sweet corn (20%). Losses of 5% occur with green beans, cabbages, potatoes, and tomatoes. Low and high losses on the basis of total material are 25 and 12% for apricots and green beans, respectively, and 58 and 85% for grapefruit and butter beans, respectively. Butter bean wastage is very high because unlike green beans, the pod becomes waste.

The wastes from canning processes include both liquids and solids. Solids are the main wastes. The liquids are less in volume and have minor concentrations of solids. An exception is potatoes, where the liquid contains high concentrations of starch. Meat processing produces much larger amounts of solid and liquid wastes than canning. Dairy processing produces little solid and liquid waste, but the liquid waste is highly concentrated. The biological oxygen demand (BOD) of liquid cheese whey wastes is about 20-fold greater than that of red meats and poultry, and about 25 times that of liquid wastes from fruit and vegetable canning.

Changes in processing methods can often lead to significant reductions in waste (Moon 1980). For example, a switch from mechanical to lye peeling can increase product recovery by 6–40%; however, the liquid waste now has a BOD about three times more than before. Use of ultrafiltration in cheese manufacture can increase yields 16–20% by retaining more protein. The resulting cheese whey has less protein and a reduced BOD. Meat wastes can be cut by 39% through the use of enzyme tenderizers.

The preceding methods increase production by raising processing yields, that is, they are prewaste technologies. Because they increase profitability and reduce waste levels, they are readily adapted by food processors. More, perhaps the most research should be directed toward developing more of these approaches in view of their benefits and the increasing need for food. Such an approach is to be preferred to one that deals with waste recovery unless wastes are inevitable.

Still, even with improved processing, the biggest problem, waste disposal, remains. Two general approaches for handling waste products are possible: recovery and disposal. At present, disposal of food wastes by anaerobic and aerobic digestion is the most common way for handling food processing wastes. Digestion systems are based upon existing sewage treatment technology. The ease of adapting such systems to food processing operations and their relative low cost have made anaerobic and aerobic digestion the method of choice (Moon 1980). Although treatment by digestion may make food wastes more suitable for environmental disposal, it does not recover any of the food value in the wastes, which can be considerable. For example, annual wastes from the U.S. tomato canning industry total about 5 million MT (wet); such wastes contain enough food energy value to feed some 2.5 million people for one year (Knorr 1983).

Methods do exist for the recovery of food or animal feed byproducts from food wastes. Increased public pressure to reduce water pollution and/or rising byproduct values could result in increased uses of these recovery techniques. Research should place emphasis on these food and feed recovery methods, not only for to increase food sources, but to decrease environmental problems. However, as mentioned previously, processing technologies to avoid wastes are preferable.

Waste recovery techniques currently available include electrodialysis, ion exchange, reverse osmosis, ultrafiltration, precipitation, single-cell protein production, and silage production (Moon 1980). The first four techniques are nondestructive recovery methods, but their operation and maintenance costs are high.

Cheese whey subjected to ultrafiltration yields a whey concentrate, containing 42% protein, which can be used in foods. Reverse osmosis has been used to recover protein from oilseed flour wastes. Electrodialysis reduces the ionic contents of protein wastes (e.g., cheese whey). High costs limit the usefulness of these approaches, but rising costs of waste treatment and protein formulation plus process improvements through research may increase their profitability.

Methods that produce limited destruction, but are less expensive, are presently more acceptable in commercial operations. The primary method of choice is precipitation to recover protein. Precipitation can be induced by the use of chemicals or heat. Chemicals include lignosulfuric acid, hydrochloric acid, polyphosphoric acid, sulfuric acid, and calcium hydroxide. However, because these chemicals must be approved as food additives, heat precipitation often is preferred. Protein products produced through chemical means are often used as animal feeds; a good example is the animal feed derived from lignosulfuric acid treatment of meat-packing effluents (Moon 1980).

A well-documented application of protein recovery through precipitation is described in detail by Knorr (1983). Potato starch effluents contain proteins, which now appear to be economically feasible to recover. After precipitation, the protein is dewatered to produce a concentrate, which is then dried. The final product is currently utilized in animal feeds, but research is underway to examine recovered potato protein for human consumption. One promising application is the partial enrichment (up to 10%) of bread with potato protein concentrate (Knorr 1983).

Another promising technique is conversion of wastes into single-cell protein (Moon 1980; Knorr 1983). Microorganisms are cultured in the wastes; they metabolize various materials, increasing their numbers. At some point the single cells are harvested through centrifugation or ultrafiltration and utilized in animal feeds. Production has been demonstrated with many of the wastes resulting from food processing.

A good example is the production of single-cell protein from the starchy wastes of potato processors (Knorr 1983). Two microorganisms are used in a symbiotic arrangement: *Endomycopsis fibuliger* degrades starches to glucose; a yeast, *Candida utilis,* then consumes the glucose. The fermentation requires about 10 hr. The yield of protein, based on starch content, is 55%. The remaining liquid waste has a BOD 90% lower than the original material and has lost at least 50% of the nitrogen and phosphorus salts. The main drawback to single-cell protein production is cost, but this restriction may drop as the efficiency of these systems is improved and the cost of animal feeds increases.

Much research is required if the reduction or reclamation of food wastes is to become part of a sustainable agriculture (Knorr 1983). The highest priority should go to developing improved food processing technology so that food yields increase and wastes decrease. Second priority should be directed at maximizing recovery of food and feed byproducts where wastes are inevitable. All this research should attempt to also reduce energy needs, process costs, and BOD demand of the wastes, yet maximize nutrient quality.

The potential return from such concerted, interdisciplinary research projects can best be illustrated with an example (Mann 1981) involving several techniques to reduce wastes in cottage cheese manufacture. Ultrafiltration can be used to recover protein from the whey wastes. The reclaimed protein is then recycled through the cottage cheese production, and the treated whey is fermented to produce methane and alcohol. These fuels are then utilized as energy sources for the cottage cheese processing equipment. This system is indicated to be economically feasible. Wastes are virtually eliminated, energy costs decreased, and less damage to water quality results.

On-Site Food Processing

Another interesting possibility is on-site food processing by family farms. Some energy savings may be possible in that transportation to the food processor would be eliminated, and possibly transportation by the farmer to a part-time job. Many family farms are a part-time operation not by choice, but by financial constraints. Income from processed foods could help to make family farms financially self-sufficient and reduce energy costs for transportation, which as will be noted later are substantial. No figures on possible energy savings exist, but the concept is under active investigation by the USDA (Walker 1982).

Researchers envision much beyond the present concept of processing and marketing: the selling of fresh produce at roadside stands, pick your own operations, and farmers' markets. Future trends for processing and marketing could include the washing, sorting, and packaging of fresh fruits and vegetables for nearby stores and restaurants; the sorting and packaging of nuts, edible or sprout seeds, dried beans, and specialty flours and meals for health food stores; and the preparation and packaging of honey, maple syrup, and cider for local stores and on-farm sales. Other possibilities are for partially processed foods, such as sprouts or apple slices ready for pies. Processed foods with limited markets that have small profit value to large processors could also be handled on the farm: sausages, cured meats, pickles, preserves, ethinic foods, and unusual cheeses.

Indeed some of these ideas already have been put into limited practice. Many are being examined by the USDA for development of appropriate food technology. Some results are apparent already. USDA scientists have invented an inexpensive means of producing and preserving maple syrup, and

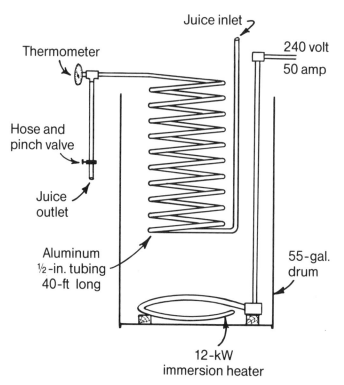

FIG. 6.2.   The cider pasteurizer unit can be built by the farmer for $200. The orchardist can now be a small-scale food processor and realize a high cash return.
Courtesy U.S. Department of Agriculture

a do-it-yourself cider pasteurizer costing about $200 (Fig. 6.2). Presently most appropriate food technology is designed for developing third world countries, but such technology could have potential on-farm use in the United States. Bates (1983) has reviewed appropriate food technology in the third world context.

In- and Out-of-Home Food Preparation

Energy used for in-home food preparation is substantial. It runs a close second to that used in food processing (Table 6.1) and accounts for 4.3% of total U.S. energy consumption. Much of the energy usage derives from re-frigerated or frozen storage, preparation in small appliances, and heating in ovens and on ranges. A large amount of energy is also expended on trans-portation to and from food stores. For example, in one study (Brown and

Batty 1976) 46–119% more energy was used in transportation by shoppers to buy a can of corn (0.454 kg) than was consumed in the nonirrigated production of the corn on the farm. These data are based on a total travel distance for shopping of 4–8 km (2.5–4.9 miles). The same study also indicated that preparation energy ranged from about the same to 45% more than that involved in production, depending on the number dishwasher cycles used per day. Based on these data, energy costs for in-home use of a can of corn were 246–360% greater than the energy expended in farm production.

Eating meals away from the home ranks third in terms of energy consumption in the U.S. food system (Table 6.1). Estimates are that over one-third of all meals are prepared outside the home (Vilstrup 1981) in conventional and fast-food restaurants, hotels and motels, coffee shops and cafeterias, and other eating facilities. Energy is needed for cooking, heating, refrigeration, air conditioning, hot water, dishwashers, lighting, and for the manufacture of disposable materials. The latter includes paper cups, plates, napkins, wrappers, and boxes; straws; and plastic knives and forks. Much of the disposable materials are used in fast-food restaurants and pose considerable challenges to energy conservationists. A good procedure for learning more efficient-energy procedures for large-scale food handling is available (Unklesbay and Unklesbay 1980).

Transportation to and from outside eating places also consumes energy. For example, Fluck and Baird (1980) point out that consumers will use 800 MJ of gasoline to eat an 8-MJ meal at a restaurant. Since transportation consumes much energy in meal preparation in and out of homes, it can be a logical place for energy conservation. Transportation is also an essential part of food processing, marketing, and distribution. Because of its widespread involvement in the food system, transportation is treated later in a separate section.

Other steps toward reducing energy usage can be common to both homes and eating establishments. These include standard insulation improvements to reduce energy use for heating and cooling, and the use of more energy-efficient appliances and larger equipment. The latter would include more energy-efficient cooking systems, especially microwave units (Mandigo and Janssen 1982), furnaces, boilers, dishwashers, air conditioners, refrigerators and freezers. Steps must also be taken to reduce the energy used in the manufacture of disposable materials.

Marketing and Distribution

Marketing and distribution of food in the United States involves a complex network of wholesale and retail outlets. About 75% of the energy used in this area is consumed in the movement of food through the wholesaler and then retailer or other commercial users of food. The energy is used for transportation, refrigeration, storage, packaging, heat, and electricity. The most important area for conserving energy is transportation, the common link throughout all components of the U.S. food system. Transportation is covered in the next section.

Other measures for reducing energy usage in the marketing and distribution of food are the use of better insulation and more energy-efficient equipment. Additional steps could include automated systems for the most efficient use of energy in warehouses and retail stores, the increased use of controlled-atmospheric storage for fruits and vegetables, and high-temperature pasteurization for milk to reduce the need for refrigeration.

Transportation

Transportation is an integral part of the U.S. food and fiber system, providing continuity between the farmer and consumer, and all intermediates in between the two. Every part of the food system is absolutely dependent upon uninterrupted transportation.

Not unexpectedly, the U.S. food and fiber system consumes large quantities of fossil fuel energy. Annual fuel requirements, based upon Barton (1980), are about 1.09×10^9 dekaliters (2.89×10^9 gal.) of diesel fuel and 1.56×10^8 dekaliters (4.11×10^8 gal.) of gasoline. The percentage distribution of this fuel is as follows: for shipment of farm inputs, 18.7%; for movement of agricultural products to food processors, 42.9%; and for delivery of food products from processors to warehouses and supermarkets, 38.4%. Trucking accounts for 77% of the total fuel used; railroads for 20%; and boats and barges for the rest. Estimates are that almost half of the trucks on major U.S. highways are carrying food and other agriculturally related products. This results from the need for timely, rapid, and efficient movement of agricultural inputs and perishable harvests (Vilstrup 1981).

A graphic example of energy consumption for transportation is provided by Brown and Batty (1976). The energy used to transport a 0.454-kg can of corn 711 km (440.8 miles) or 1600 km (992 miles) is, respectively, 51.5 and 106.4% more than the energy used for the farm production of the corn under nonirrigated conditions.

Transportation accounts for over 12% of the energy used in the U.S. food system and about 2% of total U.S. energy consumption; this is equivalent to about two-thirds of the energy needed for crop production in the United States, or nearly nine times the energy used for greenhouse heating. These amounts would certainly rate considerable attention in terms of energy conservation.

Some steps can cut energy consumption by transportation. Some can be implemented relatively quickly, such as improved scheduling and routing of shipments, and the reduction, if feasible, of wasteful empty backhauls. Other short-range changes include increased cargo efficiency through standardized modular containers, uniform packaging, and a reduction in bulk volume (compacting, cutting water content); the elimination of transportation regulations and practices that only accomplish or encourage energy waste; and instruction in energy conservation for truck drivers, truck maintainers, and dispatchers. Long-range changes include improved design of truck engines and trailers to reduce road and air friction and to increase fuel economy, and the replacement of aging or obsolete equipment by improved vehicles. These individual energy-savings steps are claimed to reduce energy needs by a few to 50% (Argonne National Laboratory 1983).

Fiber

Postproduction energy is used not only for food processing but also for the processing of fiber crops. A few analyses of energy use by these sectors have been conducted. One basic study involved the processing of cotton.

Harvested cotton requires ginning, processing or weaving, and then dyeing or finishing. Winkle *et al.* (1978) reported that these steps consume 29 kilowatt-hours (kWh) of energy for every 45.5 kg (100 lb) of baled cotton lint. Additional steps needed to manufacture shirts consume the following: 7.8 kWh/m^2 (6.5 kWh/yd^2) of cloth and 1.2 kWh/m^2 of finished shirt. Shirts manufactured from blends of cotton and polyester consume more energy.

A quick conclusion would be that energy conservation could result from the return to all cotton shirts. Such a hasty conclusion would be false and demonstrates the need for an energy analysis of the complete system. Winkle *et al.* (1978) examined energy needs further, that is, over the life time of the shirt. That analysis included energy consumption by washing, drying, and ironing. The result was that a polyester and cotton blended shirt (65/35) consumed only 62.7% as much energy over its lifetime as did a cotton shirt.

References

ARGONNE NATIONAL LABORATORY. 1983.   Truck Energy Efficiency Measures Matrix. Report of Laboratory's Center for Transportation Research and the Trucking Industry. Argonne, IL.

BARTON, J. A. 1980.   Transportation and Fuel Requirements in the Food and Fiber System. Agricultural Economics Report 414, U.S. Dept. Agric., Washington, DC.

BATES, R. P. 1983.   Appropriate food technology. *In* Sustainable Food Systems. D. Knorr (Editor). AVI Publishing Co., Westport, CT.

BROWN, S. J. and BATTY, J. C.   1976.   Energy allocation in the food system: A microscale view. Trans. ASAE *199* (4) 758–761.

CUMMING, D. B.   1980.   The development of a new blanching system. J. Can. Dietetic Assoc. *41* (1) 39–44.

FLUCK, R. C. and BAIRD, C. D.   1980.   Agricultural Energetics. AVI Publishing Co., Westport, CT.

FORWALTER, J.   1979.   Waste heat from refrigeration system provides energy for 140°F hot water, has payback of 2 years. Food Process. *40* (13) 60–61.

GRIFFITH, H. E., MALVICK, A. and ROBE, K.   1979.   External heat exchangers on retorts save TVG $59,000/yr. Food Process. *40* (5) 156–158.

HALLOWELL, E. R.   1980.   Cold and Freezer Storage Manual. 2nd ed. AVI Publishing Co., Westport, CT.

HIRST, E.   1973.   Energy Use for Food in the U.S. ORNL-NSF-EP-57. Oak Ridge Natl. Lab., Oak Ridge, Tenn.

INABA, L. K., EAKIN, D. E. and CLARK, M. A.   1981.   Energy Conservation Equipment Applications in the Food Processing Industry. U.S. Dept. of Energy, Washington, DC.

INTERNATIONAL INSTITUTE OF REFRIGERATION.   1978.   Cooling, Freezing, Storage and Transport; Biological and Technical Aspects. Paris, France.

KNORR, D.   1983.   Recycling of nutrients from food wastes. *In* Sustainable Food Systems. D. Knorr (Editor). AVI Publishing Co., Westport, CT.

LEO, M. A.   1982.   Energy conservation in citrus processing. Food Technol. *36* (5) 231–233, 244.

LEVIS, A. H., DUCOT, E. R., WEBSTER, T. F. and LEVIS, I. S.   1981.   ALINET: a model (of the U.S. food processing and distribution sector) for assessing energy conservation opportunities in food processing. *In* Agricultural Energy: Selected Papers and Abstracts from the 1980 ASAE National Energy Symposium. Amer. Soc. Agric. Eng., St. Joseph, MI.

MANDIGO, R. W. and JANSSEN, T. J.   1982.   Energy-efficient cooking systems for muscle foods. Food Technol. *36* (4) 128–133.

MANN, E. J.   1981.   Energy conservation in dairy factories. Dairy Ind. Intern. *46* (11) 11–12.

MOON, N. J.   1980.   Maximizing efficiencies in the food system: A review of alternatives for waste abatement. J. Food Protection *43*, 231–238.

MORRIS, C. E.   1979.   Soybean processing takes another step forward. Food Engin. *51* (10), 86–87.

OLABODE, H. A., STANDING, C. N. and CHAPMAN, P. A.   1977.   Total energy to produce food servings as a function of processing and marketing modes. J. Food Sci. *42*, 768–774.

PIEROTTI, A., KEELER, A. G. and FRITSH, A. J.   1977.   Energy and Food. CSPI Energy Series X, Center for Science in The Public Interest, Washington, DC.

RAO, A.   1980.   Energy consumption for refrigerated, canned and frozen snap beans and corn. J. Food Process Eng. *3* (2) 61–76.

RUTLIN, N.   1980.   Heat recovery from refrigeration plants. Food Process. Ind. *49*, 33, 35.

SINGH, R. P.   1978.   Energy accounting in food process operations. Food Technol. *32*, 40.

SINGH, R. P.   1981.   Energy-saving ideas for food processors. *In* Cutting Energy Costs (The 1980 Yearbook of Agriculture). J. Hayes (Editor). U.S. Dept. Agric., Washington, DC.

SINGH, R. P., CARROAD, P. A., CHINNAN, M. S., JACOB, N. L., and ROSE, W. W.   1980.   Energy accounting in canning tomato products. J. Food Sci. *45*, 735–739.

SWARTZ, J. B. and CARROAD, P. S.   1979.   Recycling of water in vegetable blanching. Food Technol. *33* (6) 54–59.

THOMPSON, J. F., CHINNAN, M. S., MILLER, W. M. and KNUTSON, G. D.   1981.   Energy conservation in drying of fruits in tunnel dehydrators. J. Food Process Eng. *4*, 155–169.

UNGER, S. G.   1975.   Energy utilization in the leading energy consuming food processing industries. Food Technol. *29* (12) 33–45.

UNKLESBAY, K. and UNKLESBAY, N.   1980.   Mastering basic energy concepts for effective food handling. J. Amer. Dietetic Assoc. *77* (3) 301–303.

USDA.   1979.   Handbook of Agricultural Charts. Handbook 561. U.S. Dept. Agric., Washington, DC.

VALENTIN, P.   1980.   Energy conservation studies in the beet sugar industry. The Intern. Sugar J. *82,* 303–308.

VILSTRUP, D.   1981.   Less energy, more food. *In* Cutting Energy Costs (The 1980 Yearbook of Agriculture). J. Hayes (Editor). U.S. Dept. Agric., Washington, DC.

WALKER, A.   1982.   Small-Scale Commercial Food Processing Holds Promise for Family Farms. Research News (Jan.). USDA, Agricultural Research Service, Northeastern Region.

WINKLE, T. L., EDELEANU, J., ELIZABETH, A. P. and WALKER, C. A.   1978.   Cotton versus polyester. Amer. Sci. *66,* 280–290.

7

Sustaining Resources: Soil

Farmers, with notable exceptions, have benignly neglected the future farmers and their soil reserves. Exceptions include organic farmers and other conservation-minded farmers. With others the view is toward profits now and let other generations worry about the future. In all fairness their actions, although not the best choice, are understandable. Farmers have very large capital investments in land, equipment, seeds, fertilizers, fuel, and pesticides. Their profession often does not generate high yearly profits and disaster is sometimes only a change in the weather away. Many are financially overextended and faced with credit problems. Operating so close to the margin, they grasp at short-term profits and can ill afford the long-term investment needed for sustaining the soil.

This approach has not caused any immediate problems in terms of crop productivity. Like a large bank account, constant withdrawals can provide a suitable income, but unless some deposits are made, the account will run dry someday. Indeed the account is running out: The soil bank is being emptied by erosion, the most serious problem of agriculture today.

Examples of farmers mining the soil abound. Over the last 100 years Iowa has lost 50% of its topsoil. USDA calculations are that erosion is decreasing crop productivity equivalent to the loss of 506,072 ha (1.25 million acres) of land per year. This is equal to an annual loss of 0.4% of the land under cultivation. Another way of viewing it is that the loss of 2.54 cm (1 in.) of cropland topsoil occurs every 8–10 years. Since with some soils it takes about 100 years under agricultural conditions to create an inch of topsoil from bedrock, the enormity of the problem is very obvious.

The ironic aspect is that agricultural practices exist that, if practiced, would decrease the erosion of topsoil to the rate comparable to its creation from bedrock. Soil sustainability is within reach but not achieved because to do so costs money and produces no immediate return on the investment.

On the other hand, attention to maintaining nutrient levels in the soil has been much greater. Since decreased crop productivity quickly follows reduction of soil nutrients below optimal levels, it is not surprising that farmers are more willing to use fertilizers than to use practices that prevent erosion. Sustaining of soil nutrients by fertilization is not without its problems. Exces-

116

sive fertilization has caused environmental pollution, especially with water; and too little fertilization results in decreased productivity. Fertilizer prices continue to rise, and fertilization uses up to one-third of the energy needed for crop production. Again, some existing approaches for sustaining nutrients can reduce these problems.

Sustainability of soil water has also become a problem, as our water supplies no longer appear limitless. Competition for water from nonagricultural sectors is increasing. What supplies are available can be all too readily polluted by nutrient- and pesticide-loaded runoff. Energy input into irrigation can be quite high. Again, corrective practices are available and are discussed in subsequent chapters.

The depletion of soil resources and practices suitable for sustaining soil are examined in this chapter.

Organic Matter

A certain amount of organic matter is needed in agricultural soils. Decomposition of organic matter by microorganisms releases nutrients, including trace elements, needed for crop production. While not enough for complete crop maintenance, it does help to offset fertilizer needs. The organic matter remaining after degradation, a stabilized form resistant to further microbial attack, is known as humus. Because of its cation-exchange capacity, humus is of value in terms of nutrient retention and regulated availability. Organic matter also has a buffering capacity that helps to prevent sudden pH changes resulting from the addition of fertilizers, lime, or sulfur. Organic matter also plays a role in soil structure; hence it affects aeration, water movement, and tilth. Finally it contains micropores, which help in water retention (Poincelot 1980).

The specific effect of organic matter upon soil structure arises from its contribution to the development of soil aggregates. Improvements in aggregation result in better root development and lesser amounts of energy to work the soil. A direct relationship between organic matter and the population and distribution of beneficial soil biota is also noted. Indeed, the most productive agricultural soils possess good soil structure, considerable ion-exchange capacity and water retention, and high populations of beneficial microorganisms and soil invertebrates (OTA 1982), all of which depend upon the presence of organic matter.

A definite quantitative link exists between erosion and organic matter in the Universal Soil Loss Equation (Wischmeier and Smith 1978). In the physical sense, organic matter at or near the surface improves water infiltration, while deeper organic matter improves drainage and water retention. As organic matter is lost through cultivation, water absorption and drainage deteriorate, increasing surface runoff and, subsequently, erosion of soil by water. The soil also becomes drier and more finely granulated as organic matter content declines, since decreased water retention and soil aggregation result from reductions of organic matter. Soils most susceptible to erosion by wind are dry and finely granulated.

An ideal agricultural soil can contain as much as 5% organic matter by weight. However, continuous cropping and maximal tillage exposes lower organic matter to increased aeration and microbial activity, resulting in organic matter losses by oxidation followed by increased wind and/or water erosion. The latter not only results in additional losses of organic matter but also losses of valuable topsoil. Organic matter losses can be as great as 1.5% annually. Unless the last organic matter is replaced, unfavorable changes in soil structure, increased erosion, and subsequent decreases in crop productivity will occur eventually. The addition of fertilizer alone will not suffice to maintain crop productivity over the long run. Fertilizers must be supplemented with a steady input of organic matter.

Organic matter content can also be raised above the natural level, and with some soils this may be desirable. For example, water retention and soil structure of sandy soils can be improved with organic amendments. Improvement of soil structure can result from additional organic matter in clay or silt–clay soils. These soil structure changes produce soils that have improved aeration, tilth, and water movement. The beneficial effects of organic amendments are well documented (Poincelot 1975).

Although the value of organic matter is well established, some research remains to be done (Parr and Papendick 1983). For example, the effects of agronomic wastes, such as manures and crop residues, upon tilth, soil productivity, and fertility are reasonably well known. However, the effects of others, such as various types of municipal and industrial wastes, are not well characterized. If these wastes are to become agronomically useful, they must be investigated further so that each waste can be used in the most suitable soil and cropping system to achieve maximal returns. Examples of needed information on waste interactions with various soils and crops include rates of decomposition, nutrient availability as a function of time, phytotoxicity, effects upon insects and disease problems, effects of climate, effective loading rates, and maximal loading rate. Other useful studies include economic evaluations of each waste, examination of effectiveness in reducing soil erosion, and an assessment of crop response to application methods. For example, one waste might be more effective plowed under, and another as a mulch. One waste might be more effective against erosion on a slope than another waste, but give poorer crop response on marginal soils than the other.

Fortunately, a number of practices exist that can stop the decline of soil productivity resulting from the loss of organic matter and soil erosion. Some are old practices dropped by the wayside and others are new. Organic farmers utilize a number of these practices, as do many traditional farmers. Still, as we will see later, present efforts are not enough to sustain our soil resources.

Organic Amendments

As noted already, a number of organic amendments can be used to maintain or increase the content of soil organic matter. Some of these are waste products that originate from agricultural activities such as crop production and animal husbandry; others are waste products from postproduction processing or utilization of agricultural products. These wastes have great potential as a renewable agricultural resource for maintenance of soil. They are also less costly and energy wasteful than nonwaste organic amendments. Their use will also alleviate waste disposal problems and minimize environmental pollution.

These considerations were not lost upon the federal government. The Food and Agriculture Act of 1977 (PL 95–113) directed the USDA to write a feasibility report for Congress on "the practicability, desirability, and feasibility of collecting, transporting, and placing organic wastes on land to improve soil tilth and fertility." This report (USDA 1978) was the result of factors leading to a movement toward sustainable agriculture: increased costs for energy, fertilizers, and pesticides and the problems of soil erosion and decreased soil productivity. We will now turn to discussion of various organic wastes.

Animal Manures

About 22% of the present organic wastes in the United States are manures. These include the feces and urine, but not bedding and litter, of dairy and beef cattle, horses, swine, poultry, sheep, and goats. Annual production is around 1.2 billion tons of wet manure, or about 175 million tons of dry manure. Some useful manure data is shown in Table 7.1.

The tonnage and value of nutrients shown in Table 7.1 are estimates and represent potential values. The nutrient values of manures are highly variable and depend on several factors: animal age, type of feed, bedding (type, presence or absence), waste management conditions, and storage length. The average values for nutrient composition shown in Table 7.2 were used to calculate the potential nutrient contents of the manures produced annually in the United States. These values, listed in Table 7.1, are seldom achieved because nutrients can be lost through leaching and ammonia volitalization. The latter depends upon storage and conditions during and after application. Means to maximize nutrient recovery are discussed later. The goal is to retain

Table 7.1. *Annual Animal Manure Production in the United States*

| Animal | % of total | Habitat (%)[a] | | Potential nutrients | | | | | |
| | | Confined | Unconfined | Amount (tons $\times 10^3$)[b] | | | Values ($ $\times 10^6$) | | |
				N	P	K	N	P	K
Dairy cattle	17.3	35	65	740	175	910	580	241	281
Beef cattle	58.2	75	25	2040	816	1693	1599	1124	523
Horse	13.3	31	29	512	233	373	401	321	115
Poultry	5.9	7	93	381	180	198	299	248	61
Swine	4.3	20	80	451	191	310	354	263	96
Sheep and goat	1.0	87	13	49	18	79	38	25	24
Total		61	39	4173	1613	3563	3271	2222	1100

Source: Adapted from data in USDA (1973, 1982).
[a]Values refer to percentage of manures from each animal produced in indicated habitat.
[b]Values based upon average nutrient composition of various manures listed in Table 7.2.

as much as possible of the 9.3 million tons of nutrients, valued at $6.6 billion, generated each year.

Some manure nutrients are easily recycled. Manure excreted in unconfined habitats (range and pasture) is recycled through the existing grasses and forages. This accounts for 61% of manure production with a potential nutrient value of $4 billion. The actual value is less in that some nitrogen is lost, but most of the phosphorus and potassium is not. The extent of nitrogen loss depends upon moisture and soil. On cool, wet soils at least 10% is lost after 7 days, and 50% or more is lost on warm, dry soils (USDA 1978). Taking these losses into account, the useful nutrient value is $3–$3.8 billion. The monetary value of these manures is obvious, and their energy value is equivalent to the energy contained in 0.8–1.4 billion gal. of diesel fuel.

About 39% of the total manure, with a potential value of about $2.6 billion, is produced under confined conditions. However, since only 73% is applied to agricultural land, this potential value is reduced to about $1.9 billion. The remaining nutrients (27% valued at $0.7 billion) are lost to agriculture, since the manure is treated as a waste. From the manure applied to the land, some nutrients are lost during storage and because of inefficient practices for land incorporation. About one-third of the nitrogen is lost, and lesser amounts of phosphorus and potassium. If practices for handling confined manures were improved and none wasted, the improvement in terms of energy and nutrient conservation, as well as monetary value, would be about 50% (USDA 1978).

Efficient methods of handling, storing, and applying manures do exist. Assuming their use and no manure wastage, the majority of this improvement could be realized without increased water pollution. Such management and application systems are detailed well by Gilbertson *et al.* (1979) and

Merkel (1981), so only some brief points are mentioned here. Manure piles should be covered to reduce their exposure to rain, which causes leaching of nutrients. This practice conserves nutrients and reduces water pollution. Losses of nitrogen during storage can be further reduced through compaction to make the manure anaerobic and by minimizing storage time, i.e., applying and incorporating the manure as quickly as possible into the soil. Manure stored in lagoon systems should not be held for long periods, since half or more of the nitrogen can be volatilized as ammonia.

Manure solids should be applied with rear-discharge spreaders and slurried with tank wagons or through irrigation equipment after proper dilution (Fig. 7.1). Applications should be as close to the time of plowing or disking and planting as is possible. Nitrogen losses through volatilization and leaching will be minimized with this practice; ammonia losses from surface-applied manure are substantial and placement in the root zone leads to efficient utilization and less leaching of nutrients. Ammonia losses have been found by Lauer *et al.* (1976) to range from 61 to 99% between 5 and 25 days after surface applications of manure that are not incorporated into the soil.

Rates of manure applications must be sufficient to supply reasonable levels of nutrients, yet not so excessive as to cause water pollution. Since the type of soil and other variables affect the utilization of manure, rates of application should be calculated for the sites in question. An excellent procedure for the calculation of acceptable agronomic rates and proper application times is available (Gilbertson *et al.* 1979). In general, manure application rates below 20 MT/ha (8.9 tons/acre) will give reasonable fertilizer value and little to no pollution. This figure assumes proper application and incorporation into the soil.

Excessive application of manure over time can cause problems because crops are not able to use all the nutrients, so soluble salts, especially nitrates, leach away in drainage water or through surface runoff. The latter is most likely with surface applications of manure. Water pollution results and soluble

Table 7.2. *Average Nutrient Content of Animal Manures*

	Percentage of dry weight		
Animal	N	P	K
Dairy cattle	2.4	0.6	3.0
Beef cattle	2.0	0.8	1.7
Horse	2.2	1.0	1.6
Poultry	3.7	1.7	1.9
Swine	5.9	2.5	4.1
Sheep and goat	3.0	1.1	4.8

FIG. 7.1. Liquid manure is being applied to the field with an effluent spreader. Solids can also be applied to fields in a hopper-type of spreader drawn by a tractor.
Courtesy U.S. Department of Agriculture

salts can rise to levels that reduce soil productivity, especially on irrigated land or areas with low rainfall. Manure must be applied sparingly on frozen or snow-covered land, since spring runoff can carry away nutrients and organic contaminants. Fall applications may be undesirable in some areas, as nitrate leaching can occur.

Another problem associated with excessive applications of manure is microbial pathogens, especially their possible pollution of water supplies. Examples of pathogens associated with animal wastes include *Bacillus anthracis, Clostridium, enteroviruses, Erysipelothrix, Mycobacterium,* and *Salmonella,* to name just a few. A number of variables influence the potential for pollution resulting from microbial pathogens in soils treated with organic wastes (Reddy *et al.* 1981). The first variable is the survival rate of pathogens in soil-waste systems. Such rates are influenced by several factors: waste pretreatment, feed antibiotics, soil parameters (temperature, pH, moisture, nutrients, organic matter content, soil type, competitive organisms), sunlight, and the method and timing of waste applications. A second variable is the retention of pathogens by soil particles, especially clay. Another variable is pathogen transport by leaching and runoff.

Certain general principles should be followed if one wishes to minimize problems with microbial pathogens. Avoid excessive applications of manures and use practices that minimize pollution problems associated with leaching and runoff, such as those previously discussed. Avoid winter and surface applications; incorporate manures into the soil near planting times.

Agronomic applications of manure have definite advantages beyond sustaining organic matter, according to a review by Khaleel *et al.* (1981). Manures tend to improve soil aggregation and thus stabilize soils against erosion. Another consequence of improved aggregation is a decrease in runoff volume, resulting in less pollution of water reserves by nutrients and other contaminants. Finally, moderate manure applications also appear to improve infiltration of water into the soil.

Competition for manure from other quarters is minimal (USDA 1978). Animal manures can be processed into animal feedstuffs, but are the least used of agricultural byproducts for this purpose (Owen 1980). Methane can be made from manures through anaerobic digestion, but this is seldom done in the United States (see Chapter 5), nor is pyrolysis of manure to produce fuel oil.

Adoption of more efficient management and application procedures should increase the contribution of manures to the sustainability of agriculture. However, more research is needed to further reduce the loss of nitrogen experienced even with good techniques. The savings of fertilizer and energy would be substantial if the present unrecoverable nitrogen, about 45% of the total nitrogen in fresh manures, could be retained. Other research is needed to reduce costs for the collection, storage, and application of manure and to improve odor control (USDA 1978).

Crop Residues

An important organic waste remains after crops are harvested or grazed. These crop residues include roots, chaff, stems, and leaves. Collectively, they constitute 54% of the organic wastes produced in the United States. Crop residues are an agricultural asset that can contribute to the control of soil erosion and sediment transport, the retention of nutrients through runoff reduction, and the maintenance of organic matter in the soil. In addition, they can improve several soil properties. Infiltration rates for water are increased as is water retention capacity, soil aggregation is enhanced, and crusting of soil is reduced. Finally, crop residues have some fertilizer value in that they contain nitrogen, phosphorus, potassium, calcium, and trace elements.

USDA data from 1978 updated to 1982 (Table 7.3) shows some interesting crop residue information. Three crops (field corn, soybeans, and wheat) account for 86% of the residues from 15 major crops. The 15 major crops produce more than 80% of all residues, or about 462 million tons. The amounts and value of nutrients attributable to each crop are listed in Table 7.3. These data are somewhat underestimated, since roots are excluded, minor and greenhouse crops are not considered, and most food processing wastes are not counted. Of the crop residues considered, about 70% is returned to the soil. The remaining 30% is not wasted. Most of it is fed to

Table 7.3. *Annual Crop Residue Production in the United States*

Crop	Production (tons × 10³)	Potential nutrients					
		Amount (tons × 10³)			Value ($ × 10⁶)		
		N	P	K	N	P	K
Barley	6,632	50	7	83	39	10	26
Cotton	4,469	78	10	65	61	14	20
Dry beans	630	6	1	8	4	1	2
Field corn	222,324	2,468	400	2,957	1,935	551	914
Flax	765	9	1	12	7	1	4
Oats	18,499	116	30	305	91	41	94
Peanut hay	2,567	41	3	32	32	5	10
Rice	9,294	56	8	108	44	12	33
Rye	694	4	1	5	3	1	2
Seed grass	564	5	1	11	4	1	3
Sorghum	17,523	189	26	231	148	36	71
Soybeans	51,274	1,154	113	538	904	155	166
Sugar beets	695	18	2	3	14	2	0.07
Sugar cane	32	0.3	0.04	0.4	0.2	0.05	0.1
Wheat	126,223	846	88	1,224	66	12	38
Total	462,186	5,040	691	5,582	3,352	842	1,383

Source: Adapted from data in USDA (1978, 1982).

animals and small amounts (under 4%) are sold or used as fuel. Wastage is only 2.4%, or about 1 million tons. These figures do include the bedding waste portions of manure, since they originate as crop residues. Most of the bedding wastes, nearly 80%, are contributed by the dairy industry (USDA 1978).

The total nutrient value of those residues returned to the soil is about $4.6 billion. Current practices return about two-thirds of the nutrients in these residues, so the realized value is about $3.1 billion. If crop residues were not returned to the soil, commercial fertilizer usage would have to be increased about one-third over present consumption.

The USDA (1978) estimates that prudent use of crop residues can reduce soil erosion rates of 10–25 tons/acre/year to 5 tons/acre/year. The latter figure is tolerable for maintaining soil productivity on deep productive soils. Larson *et al.* (1978) showed that if reduced or no-tillage systems were utilized in the Corn Belt, less residue was required to maintain acceptable erosion rates. Recent USDA (1985) research indicates that just covering 30% of the soil surface with crop residues can reduce erosion by 60 to 70% compared with no residue cover. This raises the possibility of increased usage of residues for other purposes.

Water conservation is also enhanced by incorporation of crop residues.

Slow infiltration can lead to increased runoff and in turn severe erosion. Water infiltration rates are increased by surface crop residues or those incorporated into soils; rates may be increased by as much as 100% with some soils. Crop residues left on the surface capture snow and increase water captured from melting snow by over 30%, and also slow evaporation on wet soils. The latter increases water storage by 10–30% (USDA 1978).

Since few crop residues currently are wasted, they already make a substantial contribution to soil sustainability. However, decreased usage, resulting from competitive uses of residues as animal feed or fuel sources, could be detrimental. This has not occurred to any great extent, since reasonably priced forages discourage the conversion of residues into feed and diversion of residues from soil would increase the need for costly fertilizers.

Some diversion might be possible, without harmful effects, where reduced or no-tillage is practiced or more efficient management practices are used. However, more research is needed to determine the minimal amounts of crop residue that can provide protection against soil erosion and still maintain reasonable levels of organic matter. Other research is needed to ascertain if excesses of crop residues are used in any areas or on certain soil types, and what organic wastes may be reasonable substitutes for crop residues.

Sewage Sludge and Septage

Currently the production of sewage sludge in the United States totals 4.5 million dry tons, which constitutes less than 1 percent of the total organic wastes produced. Septage production is about 0.8 million dry tons. The potential tonnage, nutrient value, and land use is shown in Table 7.4. The total potential nutrient value of the total combined sludge and septage is around $316 million; the amount applied to land has a nutrient value of $86 million. However, these are only potential values, since much of the nitrogen

Table 7.4. *Annual Sludge and Septage Nutrient Production and Utilization in the United States*

Source	Land use (%)	Amount (tons × 10³) N	P	K	Value ($ × 10⁶) N	P	K
Sewage sludge[a]	25	180	90	18	141	124	6
Septage[b]	40	32	13	3	27	18	1

Source: Based upon update data of USDA (1978).
[a]Assuming 4% N, 2% P, and 0.4% K.
[b]Assuming 4.4% N, 1.6% P, and 0.4% K.

is lost by volatization and some leaching. Only about 40% of the nitrogen would be available for plant uptake, so the actual nutrient value is about $216 million, of which $58 million is currently utilized.

Sludge has value both as a fertilizer and a soil conditioner (USDA 1978). Agronomic rates of nitrogen can be supplied with loading rates of 6.7–22.4 MT/ha (3–10 tons/acre); suitable loading rates for phosphorus are 2.2–9.0 MT/ha (1–4 tons/acre). Loading rates chosen on the basis of phosphorus are better for a number of reasons. Fertilizer costs for phosphorus are higher than for nitrogen and fewer problems related to scheduling and heavy metal accumulation occur at lower loading rates. The one drawback is that the improvement of soil tilth is minimal at low loading rates.

Loading rates based upon nitrogen requirements improve soil physical conditions. When soil physical conditions are optimal, better yields from crops fertilized with inorganic fertilizers are realized. For this reason significant yield increases are seen on marginal farm land when sludges are used in place of inorganic fertilizers. Corn production on good crop land using sewage sludge alone, or amended with urea, was shown to be 12 and 18% higher, respectively, than it was on land treated with ammonium nitrate or urea alone (Sims and Boswell 1980). The improvements resulting from incorporation of sludge into the soil include increased aeration, permeability, water retention and water infiltration, better aggregation, and less surface crusting. These effects vary depending upon soil texture. Increased water retention is especially noted with sandy soils amended with sludge. Clay soils show notable reductions of runoff as a result of improved water infiltration and permeability. Less compaction and increased rooting depth is also noted with clay soils (Epstein 1975; Epstein *et al.* 1976).

Composted Sewage Sludge

At present, only 25% of the available sewage sludge is applied to land. Part of the resistance to use of this organic waste results from the odors, undesirable physical properties, heavy metals, and human pathogens associated with sludge. Alleviation of these properties can occur through composting action, which has received considerable attention from the USDA.

Composting is a natural process whereby organic wastes are decomposed through the action of indigenous microorganisms. The decomposition, preferably under aerobic conditions, can produce a product suitable for agronomic use. The final product is a reasonably stable form of organic matter that can be incorporated into the soil. The methods and biochemistry of the composting process have been reviewed by Poincelot (1975).

The USDA/EPA (1980) has worked out an efficient composting process for sewage sludge, which is designated as the Beltsville aerated-pile method. Essentially sludge is mixed with wood chips in a 1:2 ratio and composted in forced-aeration windrows for 21 days. The resulting compost requires curing for 30 days, then drying and screening. The screening produces a more desirable product and removes the wood chips, which are recycled. The resulting product is suitable for agronomic use, has an acceptable odor, and is reasonably free of pathogens.

Problems with heavy metals, depending upon their original concentration, also are minimized by composting. Composts produced from mainly residential sewage sludges have low contents of heavy metals. When this compost is substituted for peat in growing media, good-quality vegetable transplants can be produced safely (Sterrett *et al* 1983). Composts derived from sewage sludges loaded with industrial wastes could pose a problem if used in the production of edible plants. However, such contaminated composts could be utilized in the production of ornamental plants.

Because some sludges from industrialized areas contain excessive amounts of heavy metals and organic chemicals, they would not produce a compost that is safe for agronomic use. These sludges could become acceptable if proper abatement and/or pretreatment processes were adopted. Some elements in excess are phytotoxic (e.g., zinc, copper, nickel) whereas others are hazardous to human health (e.g., cadmium, lead, mercury). Of special concern is cadmium, which enters the food chain more readily than lead or mercury, because it is more easily absorbed and translocated in crop plants. Mercury can pose a problem with mushrooms. Guidance in this area is available (USDA/EPA 1980; Walker 1980). The Environmental Protection Agency has yet to issue formal regulations in this area, but undoubtedly will in the future.

Another cause for concern involves the presence of human pathogens in compost. Based upon past research, the heat produced during the composting process can destroy pathogens (Poincelot 1975). Temperatures must exceed 55°C (131°F) throughout the pile (especially at surfaces and edges) for 18–21 days. However, in view of the pathogenic nature of sewage sludge and the need to protect public health, some standard of quality control would be warranted with sewage sludge composts.

Burge *et al.* (1981) have proposed using the heat-resistant bacteriophage f2 as an index to pathogen destruction during composting operations. Their recommendation is that 15 logarithms of inactivation is a reasonable precaution for single pile sewage sludge compost. Such a result might be achieved if the compost reached 70°C (158°F) and had not dropped below 50°C (122°F) within the next 10 days. Two precautions must be noted. First, these temperatures must be reached near the surface and edges. Secondly, the authors feel additional research is needed to completely establish confidence in the safety of this technique and to determine the temperature requirements for extended piles.

Another health concern about composted sewage sludge deals with aerosolization of indigenous microorganisms when the working or finished compost is handled. Finished compost can be quickly repopulated by organisms that formed spores as a result of high temperatures, or by airborne microbes. Organisms of concern include the thermophilic and thermotolerant actinomycetes, especially those involved in causing farmer's lung, and *Aspergillus fumigatus,* a fungus involved in allergic reactions and secondary infections.

These aspects were examined by Millner *et al.* (1980) and Millner (1982). Such microorganisms were certainly present in sewage sludge compost and became airborne during handling operations. It appears that microbial emissions and associated health problems can be limited if allergic-prone individuals are not employed to work the compost, mechanical agitation is minimized, and dusty operations are enclosed. While no health problems were noted with compost workers, more research is needed in this area to increase confidence in the health aspects of compost processing.

The finished compost is lower in nutrients than the original sewage sludge. The typical nutrient content of compost from undigested sewage sludge is 1.3% N, 1.5% P, and 0.2% K (Parr 1982); for compost from digested sludge, the respective values are 0.9, 1.0, and 0.1% (USDA/EPA 1980). Other essential nutrients, such as calcium, iron, zinc, copper, and manganese, are present in composted sewage sludge. However, this material cannot be labeled as a fertilizer; it must be regarded as a soil conditioner needing supplementation with fertilizer. Nonetheless, the benefits of composted sewage sludge as a soil conditioner are substantial (USDA 1978; Hornick *et al.* 1979) and similar to those discussed previously in the section on organic matter.

The actual nutrient release from composted sewage sludge is gradual, thus minimizing losses through leaching. Parr (1982) has presented nutrient release figures for sewage sludge compost. All of the potassium, 60% of the phosphorus, and 10% of the nitrogen is released in the first year. Of the remaining nitrogen, 5% is released the second year, and 2% annually thereafter. Of the remaining phosphorus, 5% is released annually after the first year.

Composted sewage sludge also has some value as an amendment for controlling plant diseases present in the soil (Lumsden *et al.* 1981). This control occurs through disease repression, not pathogen destruction. Diseases repressed by compost in field studies include pea and cotton damping off (*Phythium* and *Sclerotinia,* respectively) and leaf drop of lettuce (*Sclerotinia*). In greenhouse studies other diseases were identified that may be controlled, and some that did not show any repression also were noted. One very interesting result was the synergistic effect observed with combined compost and fungicides (thiram and pentachloronitrobenzene). The effect seemed to persist into the second year.

Agronomic and horticultural uses for sewage sludge compost are numerous (Hornick *et al.* 1979; Parr 1982). Applications include turfgrass production, vegetable production, field crop and forage production, nursery and ornamental production, potting mixes, reclamation of disturbed lands, and improvement of marginal or very poor soils. Results in terms of improving crop yields have been especially noteworthy with marginal lands (Hornick 1982; Tester and Parr 1983).

The economic value of composted sewage sludge is difficult to determine for two reasons: Insufficient data is available on yield responses of crops to sludge compost under various climatic and agronomic conditions; and assigning an economic value to the soil improvement, erosion reduction, and disease repression resulting from compost applications is difficult. Barbarika *et al.* (1980), acknowledging these limitations, assigned a nutrient-related value of $20/ton less $1.50/ton applied. Thus, the value at rates of 3 ton/acre would be $15.50/ton. The deduction for increasing application rates results from the diminishing yield response observed with increased rates. Higher application rates probably do provide more benefits than lower rates in terms of soil properties and residual effects, but by how much is uncertain at present.

Based upon the present data, it would appear that increased use of sludge and sludge composts is very feasible. However, increased use would contribute only a small part toward the sustainability of agriculture. Even if all available sludge were used in agriculture, less than 1% of the U.S. cropland would be needed for application. Its fertilizer value would also represent less than 1% of that of fertilizers currently used. Nonetheless, the probability of increased use on cropland is higher for composted sludge than for animal manures or crop residues. Currently, 90% of manures is used and 70% of the crop residues, whereas only 23% of the available sewage sludge and septage is used. Usage of sludge may reach 60% by 1990, especially since land applications appear to be cost competitive with alternate disposal methods (USDA 1978).

Competitive uses of sewage sludge are presently minimal. Increased methane production by anaerobic digestion of sewage sludge is not competitive, as the anaerobically digested sludge is acceptable for land or composting applications. Self-supporting combustion of sludge may be a future competitor, depending upon technical developments, since the fuel value of sludge is estimated at 4.6 million Btu/dry ton.

Needed research related to composted sewage sludge has been surveyed in detail by Parr (1982); only a brief mention is possible here. More long-term studies are required on heavy metal accumulation and movement, interactions between toxic organic chemicals in sludge with soils and crops, and assessment of health risks associated with human pathogens found in compost. Other long-term studies are needed on the economic returns of improvements in soil properties and yield responses under various agronomic and climatic conditions. The benefits of disease repression and synergistic effects of sludge compost and fungicides would certainly warrant increased research activity. The amount of fertilizer required with sludge compost, based upon different crop and soil management systems, also needs investigation. Finally, the most effective management system for reclamation of disturbed lands and restoration of marginal land needs to be ascertained.

Other Organic Wastes

Several other activities generate organic wastes that might have agricultural uses. The contribution of each to the present total of organic wastes in the United States is as follows: food processing, 0.4%; industrial operations, 1.0%; logging and wood manufacturing, 4.5%; and municipalities, 18.1%. Land use of these wastes are estimated to be 13, 3, 5, and 1%, respectively (USDA 1978).

Food Processing Wastes

Of the wastes originating in industries that process fruits, vegetables, seafood, sugar, fats and oils, and dairy products, 89% comes from the processing of fruits and vegetables. The annual production of wastes from fruit and vegetable processing is about 3.2 million dry tons.

About 81% of food processing wastes is used for animal feed and 3% for the manufacture of byproducts. The remaining 16% is available for land application. The estimated tonnage of available nutrients is 5000, 900, and 7000 tons of nitrogen, phosphorus, and potassium, respectively, valued at $3.9, $1.2, and $2.2 million, respectively. The actual land use of these materials is difficult to determine, since data are limited. However, it appears that most wastes not used for animal feed or byproducts are used on land.

Increased land use is unlikely, since these wastes have higher value as animal feed. Diversion of these wastes from animal feed to land application, economics aside, would not contribute much to agricultural sustainability because the lost animal feed would most likely be replaced by some other more energy costly feed derived from cropland.

A decrease in land application of food processing wastes may be more likely, since the use of these wastes in animal feeds and the manufacture of byproducts is increasingly more profitable to food processors. More efficient processes to reduce waste (see Chapter 6) will likely decrease land use further.

Industrial Organic Wastes

Many industries generate organic wastes. These include the fermentation industries (alcoholic and pharmaceutical); leather tanning; meatpacking and related industries; milling; organic fiber, paper and, allied product manufacturing; and petroleum refining and related industries. The tonnage of these wastes is difficult to estimate, since adequate records are unavailable. However, it is estimated that 12 million tons of organic wastes suitable for land application will be available in 1985 (USDA 1978). The majority of organic industrial wastes, probably 95%, remain unsuitable for land use.

Most of the wastes from the fermentation industries, meatpacker and related industries, and milling industries go into animal feed. Diversion of these wastes to land use is not likely on the basis of profit and other considerations. Most of the remaining industrial organic wastes are unsuited to agricultural use because they contain heavy metals or hazardous or phytotoxic chemicals and/or because they have very high carbon to nitrogen ratios, extremely low nutrient content, or high soluble salts content. One approach to reducing the environmental and agricultural danger of these wastes would be to compost them alone or in combination with other safer wastes (Willson *et al.* 1982). An excellent examination of the problems found with hazardous industrial wastes in terms of land treatment is available (Parr *et al.* 1983).

Increased use of industrial wastes is likely to be low, and dependent upon future developments. Improvements might include industrial pretreatments, better characterization of each waste, and research to evaluate agricultural suitability of composted industrial wastes.

Logging and Wood Manufacturing Wastes

Wastes resulting from logging operations are limbs, foliage, and commercially undesirable wood. The tonnage of such waste is about 26 million. The amount of these wastes removed for local use as firewood is unknown, but undoubtedly rising in view of increasing energy costs. Costs for collection and transportation are such that these materials are not presently applied to agricultural land. Increased future use is unlikely, unless research on whole-tree harvesting and utilization results in an economical process. One area needing research is assessment of the effects on forest ecosystems of removing extensive nutrients in the form of small twigs and foliage.

Wood manufacturing wastes include bark, chips, sawdust, and assorted trimmings. Annual U.S. production is on the order of 70 million tons, of which 85% is used for pulp and paper products, various wood products, and chemical production. Some of the remaining 15% is used in horticulture, primarily in potting soil mixes and as mulches. As a rule, bark or sawdust is composted and used in container media, and wood or bark chips and sawdust can be employed as mulches. The composting of bark and sawdust container media was reviewed by Bunt (1976) and Hoitink and Poole (1980).

Increased use of wood wastes—an inexpensive, renewable resource—in horticultural growth media and mulches is likely. Peat moss, the traditional organic amendment for growth media is becoming more expensive, harder to obtain, and increasingly used as a fuel. No danger exists that the increased use will overrun the present supply.

Little increased usage of wood wastes for land application is probable. Nutrient contents are extremely low, in fact nitrogen must be added to prevent robbing of soil nitrogen during decomposition. Transportation costs are also prohibitive (USDA 1978). More efficient use of wood and increasing demand for wood wastes as fuel, pulp, and chemical sources are likely to further decrease the impetus for land use of wood wastes. The one exception is the use of wood wastes in the reclamation and revegetation of disturbed land located near sources of wood wastes.

Municipal Wastes

Annual U.S. production of municipal wastes is about 200 million tons; 70% of this consists of biodegradable organics. However, little, if any, is applied to land as an organic amendment for several reasons. Municipal wastes contain large bulky items, which must be removed, and other items that are nonbiodegradable. Shredding is required before land application or compost use. If smaller nondegradable materials are left in, they can interfere with shredders and machines used to incorporate the wastes into the soil. In addition nutrient contents of municipal wastes are low and carbon: nitrogen ratios too high for agricultural use, unless amended.

One possible use in horticulture is composting of municipal refuse with sewage sludge. The compost can be used for preparation of growth media for ornamentals and as a mulch. Economics do not justify this approach presently. The most likely use of municipal refuse may be for the generation of energy once appropriate technology is developed (Sanderson 1980).

Soil Erosion

Erosion of soil is not a phenomenon unique to agriculture, but a natural process that occurs even on land with permanent vegetation. What is unique to agriculture is the alarming rate of soil erosion, a rate exacerbated by present agricultural practices that accelerate rather than minimize the rates of erosion inherent in land utilized for crops. Present average rates of erosion usually exceed the average rate of soil formation by 10:1, causing a serious decrease in topsoil volume (Larson 1981). This leads to a gradual loss in soil productivity, which if left uncorrected, results in a subsoil having little value for agriculture. Some valuable reviews on soil erosion include those of Williams *et al.* (1980) and CAST (1982).

Annual rates of soil formation range from 1 to 5 tons/acre (2.2 to 11.2

MT/ha). Expressed another way, to create 1 in. (2.54 cm) of topsoil from the upper subsoil of well-managed, productive cropland takes 30 years. However, erosion reduces the topsoil and subsequently the depth of the root zone. Root zone formation derives from much deeper in the subsoil, including unconsolidated materials and rock. Here the annual rate of renewal is estimated to be 0.5 ton/acre (1.12 MT/ha) or less; it takes about 100 years to create an inch of topsoil from such materials. Erosion conditions are such that the USDA/CEQ (1980) estimates that 34% of American cropland is undergoing a decline in soil productivity. This dangerous state of agricultural affairs is not unique to the United States. Some 20–33% of foreign cropland is also experiencing a decrease in soil productivity (Brown 1981).

The average annual rate of water-caused erosion on non-federal land in the United States is 5 billion tons, based on 1977 USDA data. Preliminary results from a 1982 USDA survey are similar. Wind erosion rates are 1.5 billion tons/year in the ten Great Plains states; national data are not available for wind erosion. A breakdown of these rates by type of land usage is given in Table 7.5. The majority of agricultural erosion by water is from sheet (removal of essentially uniform soil layer) and rill (small channels) erosion of cropland, pastureland, rangeland, and forest. These processes account for the erosion of 413 million acres of cropland, 414 million acres of non-federal rangeland, and 134 million acres of pastureland (Fig. 7.2).

Some areas of the country are at greater risk than others as seen from the extremes in Table 7.5 and Fig. 7.3. These high risk areas are designated in

Table 7.5. *Sources and Rates of Soil Erosion in the United States*

Source	% of total	Annual average rate (tons/acre)[a]	Extreme rate		
			Highest (tons/acre)[a]	% of land	% of erosion
Erosion by water					
Roads and construction	5	—	—	—	—
Gullies	6	—	—	—	—
Streambanks	11	—	—	—	—
Cropland	38	4.7	30	2	25
Rangeland	8	2.8	5.0	12	57
Pastureland	3	2.6	5.0	11	50
Forests	29	1.2	—	—	—
Grazed forests	—	4.2	—	—	—
Erosion by wind					
Rangeland	45	2.0	14	3	31
Cropland	55	5.3	14	9	53

Source: Adapted from data in OTA (1982).
[a]Multiply values by 2.2 to convert to metric tons per hectare.

FIG. 7.2.   a—Rill erosion of rangeland in the early stages (shoestring erosion). b—Advanced rill erosion (gully erosion) of a cornfield. c—Critical sheet erosion on summer fallowed land in the Palouse Basin.

Courtesy U.S. Department of Agriculture/Soil Conservation Service

Table 7.6. Other areas of high risk, but affecting considerably smaller acreage, include regions of Colorado, Wyoming, and Kansas with wind erosion rates as high as 14 tons/acre (31.4 MT/ha) (OTA 1982).

Average and extreme rates of erosion can be put into perspective by comparing them to the rates of soil formation described already (OTA 1982). Such data is used by the Soil Conservation Service to assign soil loss tolerances, called T-values, to various soils. Rates below 1 ton/acre/year (2.7 MT/ha/year) are not realistically obtainable. T-values range from 1 to 5 tons/acre/year (2.2 to 11.2 MT/ha/year). The lower rate is for shallow soils with unfavorable subsoils and parent materials that hinder root growth; soils that have suffered heavy erosion also are assigned low T-values. The maximal value is assigned to deep, productive soils having good drainage and aeration.

About 60% of the soil types have been assigned the maximal T-value of 5 tons per acre per year. However, in view of the 10-fold lower rate of root zone soil formation, this value may be too high and may not assure the sustainability of the soil. The scientific issue is clouded further by the fact that T-

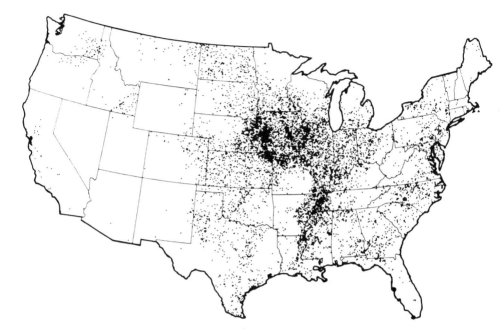

FIG. 7.3. Each dot represents 250,000 tons of soil lost annually by sheet and rill erosion. Areas of special concern include the Corn Belt, Delta States, and west Tennesee.
Courtesy U.S. Department of Agriculture

values are not established from a research base, but are determined by a collective judgment. Long-range research is needed to determine the relevance of present T-values for sustaining soil productivity with present or modified agricultural practices. An alternative approach, soil vulnerability curves (plot of productivity index vs. soil removal), has been proposed by Larson *et al.* (1983).

An examination of Tables 7.5 and 7.6 shows that some rates of erosion exceed the recommended T-values by a considerable margin. Those values that do not are acceptable, assuming the reliability of the T-value of 5 tons/acre/year. Even these lower rates of erosion may prove unacceptable over the long range.

Erosion and Crop Yields

Another way to put erosion rates into perspective is to examine their effect upon crop yields. Research in this area is not as extensive as might be expected. The impetus for such research is lacking, since crop yields continued to be adequate and profitable. Further, such research requires many years of data because the decrease in crop productivity as a function of erosion is a slow process. The cost of such long-term projects is high, especially in view of the fact that climate affects erosion and crop productivity, thereby necessitating regional studies.

However, some studies exist. For each 2.54 cm (1 in.) of topsoil eroded from cropland having a base of 30.5 cm (12 in.), an average decrease of 10 bu/ha of corn is noted (Pimentel *et al.* 1976). Analyses of yield data by Hagan and Dyke (1980) with eroded versus noneroded sloping soil in the Corn Belt showed corn yield reductions of 7.5 bu/ha (3 bu/acre) for each 2.54 cm of lost topsoil. Yield decreases become larger as erosion changes soils from slightly to severely eroded. McCormack and Larson (1980) showed such changes brought about yield decreases of 56.8 bu/ha (23 bu/acre) and 1.1 tons/acre (3 MT/ha) for oats and hay, respectively.

Other studies in western Tennessee (Buntley and Bell 1976) demonstrated crop yield differences between eroded and noneroded soils. Corn yields were reduced by 5–13% on moderately eroded soils of various types, and by 14–44% on severely eroded soils. Yields of other crops showed similar reductions on moderately and severely eroded soils, as follows: soybeans, 10–33% (moderate erosion) and 20–47% (severe erosion); wheat, 4–13% and 7–25%; cotton, 2.8–8% and 11.3–34%; and fescue, 0–17.5% and 4.8–20%. A recent USDA (1985) study shows severe erosion can cut soybean yields in half.

Not all eroded soils result in decreased crop yields. If the topsoil is es-

Table 7.6.   *Areas of Excessive Erosion in the United States*

Location	Type of erosion	Annual rate (tons/acre)[a]
Aroostock County, Maine	Water	19
Corn Belt States		
Illinois	Water	6.8
Iowa	Water	10.0
Missouri	Water	12.0
Hawaii	Water	14.2
Palouse Basin[b]	Water	50–100
Puerto Rico	Water	50
Southern High Plains[c]	Wind	20–50
Southern Mississippi Valley		
Mississippi	Water	11
Tennessee	Water	17
Texas Blackland Prairie	Water	10–20
Virgin Islands	Water	50

Source: Adapted from data in USDA (1980) and OTA (1982).
[a]Multiply values by 2.2 to convert to metric tons per hectare.
[b]Washington and Idaho border.
[c]Western Texas and eastern New Mexico.

pecially thick, little change in crop yield is noted over the short term (OTA 1982). Such results have been observed in western Iowa, even when up to 2.1 m (7 ft) of loess soil were removed.

Status of Research on Erosion and Soil Productivity

The mechanism causing decreased crop yields on eroded soils is not fully understood. However, some factors have been identified. Organic matter is lost with the eroded soil, thus the water retention and nutrient retention and release properties of the remaining soil is reduced. With some soils, such as those in the Corn Belt, yields can be almost restored by increased applications of water and nutrients. However, this does not work on eroded soils of the Southeast. These soils are very low to deficient in clay, so they lack the nutrient retention and release properties of both clay and organic matter. On eroded soils of the Corn Belt, the clay maintains some nutrient storage and release capacity, which is sufficient with the additional fertilizer.

Research is needed on whether continued applications of chemical fertilizers can maintain crop yields on eroding soil over the long term, and whether it will be economically feasible. The effect of continued losses on water retention and increased costs for water also need to be examined. Essentially, the long-term effect of erosion on the productivity of various soils in differing climates needs to be addressed. Recently the USDA established the National Soil Erosion–Soil Productivity Research Planning Committee to consider these questions. Hopefully, federal funding will be sufficient and of the necessary duration to support the efforts of this valuable committee to their conclusion.

Economics of Erosion Control

The ultimate price for failure to control erosion is high: failure of agricultural systems, followed by starvation and the toppling of a nation. An unlikely scenario, you think, but perhaps not so far-fetched. The once powerful Mayan civilization probably disappeared because of not controlling soil erosion (Brown 1981). Mesopotamia once supported 25 million people, but the area today (Iraq) carries a population of only 4 million. Soil erosion played a considerable part in this decline (Troeh et al. 1980).

Studies on the cost of erosion control exist. Rosenberry et al. (1980) calculated costs for not correcting erosion in terms of reduced yields and increased costs of additional fertilizer and energy. Their conclusion was that the short-term cost of controlling erosion was three times that resulting from the savings benefit.

Research on the actual dollar cost is limited. CAST (1982) estimates that the nutrients in the 2 billion tons of soil eroded from cropland annually are worth $8 billion. However, Larson *et al.* (1983) place the value of lost nutrients at $1.1 billion. Additional costs would include those of dredging sediment and additional water treatment, which can be conservatively estimated at $135 million (OTA 1982).

The national cost for reducing soil erosion to a reasonable level has been determined by the Soil Conservation Service (USDA 1980). An annual cost of $2 billion for the next 50 years would be required to implement their proposals. This would be in addition to the existing budget appropriations for conservation. Funding is not likely to reach these levels in the near future, given the budgetary restrictions of contemporary government and beleaguered farmers.

Basically the plan called for maintenance of existing conservation programs and implementation of new conservation programs for some 64 million ha (158 million acres) of our 167 million ha (413 million acres) cropland base. Suggested conservation measures were minimum tillage, strip cropping, contour plowing, and terraces. The adoption of this program would affect farmers and consumers, as costs would be driven upward for both. The present "do nothing" approach, however, also will drive food costs up over the long run as soil productivity declines.

Currently the major obstacle to a national effort to control soil erosion and correct its effects is that no one wants to pay the entire bill—not farmers, the national government, or the public as taxpayers. But, soil erosion is a time bomb ticking away for U.S. agriculture, and some solution must be found. An equitable arrangement for sharing the costs of erosion correction on a massive, crash basis must be arranged between the farming community and federal/state governments. One incentive might be tax credits for participating farmers.

The USDA (1981) and Congress (OTA 1982) appear close to recommending this shared expense approach, including the possibility of tax credits. The suggested federal funding share for this program to control erosion on those croplands undergoing the most severe erosion would be about $1 billion over 5 years. Public support seems to exist for shared funding, especially since major responsibility for implementation would be at the state and local level.

Unhappily public support does not exist for the massive funding program needed to correct the majority of erosion. The USDA has shown the magnitude of the problem, the necessity and costs of correcting the problem, and even proof of long-range profitability. The one element lacking is the political support necessary to implement the program. To generate such public support is a most difficult task among a mostly urban public far removed from farming and highly resistant to anything that will raise taxes.

Public awareness of erosion will increase, of course, if food costs and scarcities escalate dramatically as agricultural productivity declines. Hopefully, such a drastic stimulus to public awareness will not be necessary. A carefully planned public information program should be mounted. Once the public becomes aware of the problem and the impending disaster of failure to correct it, and is convinced of long-range economic benefits, we should see the awakening of public and political support.

Another mechanism may provide some incentive for farmers to invest in practices to control soil erosion. As soil productivity decreases, the value of the land decreases, thus providing investment motivation. This assumes the land will remain in agricultural use, as soil productivity is not important in determing the value of land that lies in the path of surburban development. Another mitigating factor is that of artifically high values for crops as a result of government price support, which in turn raises land values. If these factors become additive, the incentive for erosion control could disappear in the short run.

Practices to Control Erosion of Croplands

The erosion of cropland varies considerably, since so many variables are involved. However, some croplands are more likely than others to be targeted for erosion control. Cropland used for row crops or small grains erodes at a rate twice that of other cropland. Since this higher-eroding cropland makes up three-fourths of the total cropland, it would seem a likely target. However, in view of costs, a more conservative target would be the 9.7 million ha (24 million acres), 6% of the nation's cropland, on which 43% of total sheet and rill erosion occurs.

Practices to control erosion while maintaining productivity can be either mechanical or biological. One thing these systems must be is resource-sparing. Some of these technologies are already being adopted rapidly, such as no-tillage farming; others are still in the imaginations or laboratories of the genetic engineer.

Whichever practice is chosen, the principle of *caveat emptor* applies. The choice must ultimately be based on site specificity and the awareness that resource-sparing technologies require much more sophisticated management than resource-consuming technologies. In some instances trade-offs may be required. New attitudes, improved education, and plain old time will be needed for the implementation of new agricultural technologies. Unfortunately, it will be a race against time, since irreversible damage to cropland productivity has already commenced.

Tillage Variations

Reduced (conservation) tillage and no-till systems were discussed in terms of energy conservation in Chapter 3. Although fuel savings with these tillage systems are considerable in comparison with conventional tillage (i.e., spring or fall moldboard plowing followed by seed finishing with a disk or harrow), their main advantage is the reduction of erosion. Studies indicate that erosion associated with conventional tillage can be reduced 50–90% by a switch to conservation tillage (Phillips *et al.* 1980; Crosson 1981). An 11-year USDA (1985) study shows that conservation tillage cuts soil erosion by 70% compared to conventional tillage. The success of this tillage is attributed to the quantity of surface crop residues left behind after plowing. These residues improve moisture retention and act as a barrier against wind erosion and erosion from surface runoff.

Several tillage implements for conservation tillage are available, but the most widely used is the chisel plow. Other implements include subsoilers, disks, cultivators, mulch spreaders, strip rotary tillers, and no-till planters. Often a regional preference is noted. For example, subsoilers are frequently seen on the southern coastal plain and no-till planters in western Iowa, eastern Nebraska, and South Dakota.

These implements have been adapted to five basic methods of conservation tillage. One of these is chisel planting (Fig. 7.4). Here the bed is prepared with a chisel plow which leaves crop residues in the top 2 in. and on the surface of the soil. Planting can be carried out at the same time as plowing or later. Disk planting is very similar, except the seedbed is prepared by disking

FIG. 7.4. A chisel plow and a stalk chopper help keep crop residue near the surface, which reduces wind and water erosion.
Courtesy U.S. Department of Agriculture/Soil Conservation Service

the soil. With till-plant, plowing and planting is done in one operation with crop residues mixed into the soil surface between rows. Strip tillage also involves one step for plowing and planting of strips, with undisturbed crop residues left in place between the strips. The least disturbance is with no-till, where only the immediate row is disturbed for planting by slotting or slicing through the undisturbed crop residues. Weed control for all methods involves herbicide applications, crop rotations, and plant competition.

Data concerning the number of hectares under conservation tillage and no-till exists, but is variable depending upon the source of data. The hectares under conservation tillage, including no-till, rose from either 11.9 or 17.8 million ha (29.5 or 44.0 million acres) in 1973 to 22.3 or 32.0 million ha (55.0 or 79.2 million acres) in 1979. The first and second numbers in each pair are from USDA and private sector data, respectively (OTA 1982). The increase in no-till specifically was from 2 million ha (4.9 million acres) in 1973 to 2.9 million ha (7.1 million acres) in 1980. In 1983 the no-till cropland was about 3.7% of U.S. cultivated cropland and various forms of conservation tillage were practiced on about 30% (Greenhouse 1984). The cropland under conservation tillage is projected to rise to 70% by 2000 according to OTA (1982). In Kentucky alone, as of 1981, some 365,000 ha (901,550 acres) of no-tillage corn and soybeans were planted. This is roughly one-fifth and one-third of the corn and soybean crops, respectively (Frye *et al.* 1983).

The incentive for future adoption is present. Besides erosion prevention and fuel savings, which have been discussed, another incentive is reduced labor. Fewer trips across the field are needed with conservation tillage and no-till. This can result in as much as a threefold productivity increase (Triplett and Van Doren 1977). The labor savings has tangible benefits for the farmer: More land may be planted to crops or multiple cropping may become possible, and the optimal timing for tillage, seed germination, and weed control may be more closely approached in practice.

Machinery costs and maintenance costs can be lower for conservation tillage and no-till systems. Equipment for these systems is usually smaller and less powerful than that for conventional tillage. This advantage exists only for farms devoted entirely to conservation tillage or no-till. If conventional and conservation tillage is used on the same farm, the need for two types of equipment actually means higher costs.

A particularly attractive benefit of conservation tillage and no-till systems is the potential for increased crop production. This increase can come from multiple cropping and the expansion of row crops to sloping land.

Conservation tillage and no-till lead to a savings in time and moisture, which can compensate for climatic shortcomings that normally preclude double cropping in some locations. The usual double-cropping combinations used with conservation tillage and no-till are wheat or other small grains followed by soybeans, corn, millet, and sorghum. The feasibility of double-cropping is demonstrated by estimates that 75% of the no-till soybeans in 1980 were double-cropped. Triple sequences, mostly in the Southeast, include barley, sweet corn, and soybeans; barley, corn, and soybeans; and barley, corn, and snapbeans (OTA 1982).

The usual practice with sloping land, even with highly productive soils, is to plant forage crops as a conservation measure. However, a no-till system can reduce erosion 100-fold, making the planting of row crops feasible on sloping land as steep as 18% (Triplett and Van Doren 1977; USDA 1985). Some concern about the substitution of row crops for hay and pasture on sloping lands has been expressed by Larson et al. in 1983. Since these lands are the most susceptible to erosion, the selection of a management system to control erosion is especially critical.

Some other new forms of tillage, besides the preceding conservation and no-tillage practices, do exist. However, less research and even less attention has focused on them. Still these forms are especially attractive, in that they are especially suitable for erosion control on the most susceptible cropland, slopes. These practices, slot mulching and basin tillage, might be a better or alternate choice over no-tillage, if slopes planted with forage crops are to be replaced by row crops.

Slot mulching is a variation of vertical mulching (Saxton et al. 1981; Parr and Papendick 1983). The procedure with the latter is to incorporate organic matter into channels that follow the slope's contours. The channel is 30–40 cm (12–16 in.) deep and 10–12 cm (4–5 in.) wide at the surface. The channels are placed at 5 to 10 m intervals (16–33 ft) by tractor-driven machines. The slot mulch method utilizes a machine that collects and compresses crop residues into vertical channels or "slots," which are 15–20 cm (6–8 in.) deep and only 3–4 cm (1.2–1.6 in.) wide at the surface. The effectiveness of slot mulching to halt soil erosion has been demonstrated on steeply sloping land planted to winter wheat in eastern Washington (USDA 1985). Both techniques result in trapping of runoff water, thus reducing erosion and improving water conservation. Root penetration is also enhanced.

Another method, basin tillage, has been shown to reduce erosion substantially on slopes planted to potatoes in Idaho (Anon. 1984). The studies, conducted by T. S. Longley of Idaho University's Agricultural Research and Extension Center (Aberdeen), utilized basin tillers. These machines can be run separately or attached to existing tillage equipment. Basin tillers place small, water-holding basins in the soil, which trap runoff.

The success of erosion control with no-till systems depends upon the mulch cover. This cover can arise from residues of the harvested crop or a winter cover crop. However, maximal results can be derived from the combined effects of both (Frye et al. 1983), which increase the mulch density year-round.

The living cover crop stabilizes the soil against erosion during the winter and early spring, while the combined residues protect it other times. The cover crop can also decrease leaching of nutrients during the winter and early spring. The combined residues are effective in maintaining organic matter. More research is needed to reduce costs and maximize management efficiency.

The costs of the winter cover crop can be partially offset by using a legume, which will decrease the need of following crops for nitrogen. The aspects of legume choice and nitrogen savings were discussed in Chapter 3.

Conservation Tillage Problems. The foregoing discussion would not be complete without an examination of the problems associated with conservation tillage and no-till. The chief concern with reduced tillage is weed control. Even with the use of herbicides, a problem exists with perennial weeds. Perennial weeds must be controlled by attack at the root system, where tillage works effectively and herbicides do not. Some weeds are not easily controlled with available herbicides. In some cases periodic, mechanical weed control may be required every 4 or 5 years.

Another problem may be increased pests and diseases, both of which are encouraged by the crop residues remaining after conservation tillage and no-till. Residues on the surface provide a habitat suitable for the growth of pests and a growth substrate for plant pathogens. Control with pesticides, such as rodenticides or insecticides, is possible. Resistant varieties of plants are a solution preferable to more costly fungicides.

Recently (USDA 1985) stem canker disease with no-till soybeans has possibly been linked to surface residues. Another USDA (1985) study indicates that no-till wheat and barley may suffer reduced yields from chemical toxins (acetic and butyric acids) produced in the decaying residues. The former problem can be solved by conversion to double cropping of wheat followed by soybeans. The latter may be resolved by moving the residues aside during seeding operations. A feasibility study is in progress.

In some respects reduced tillage may pose more of a pollution problem than conventional tillage. Even though the volume of runoff and sediment is reduced under conservation tillage and no-till, sediment-associated pesticides may be greater. The reason for this is that the higher amounts of pesticides used with these tillage systems can lead to increased wash-off of pesticides from both plants and crop residues, which in turn leads to higher concentrations of pesticides in the runoff. The question is whether the reduced volume of runoff and its sediment content offsets the increase in pesticide concentration. It appears that the offset is not great enough, and that increased pollution is associated with reduced tillage (Crosson 1981). Whether the trade-off of excellent erosion control is worth more than the price of increased pesticide problems cannot be answered with present data

(OTA 1982), but rough analyses seem to imply that the increased erosion control outweighs the pollution risk.

Another possible problem noted by Larson *et al.* (1983) may arise from the fact that conservation tillage and no-till are so successful in reducing erosion. As pointed out previously, the ability to control erosion is leading to the replacement of hay and forage crops on slopes by row crops. However, the use of conventional tillage every 4 or 5 years may be required for the control of perennial weeds. This will expose the slopes to extensive erosion that year, a hazard that did not exist with pasture or forages. The loss of forage and pasture crops could increase pressures on the remaining pasture and western rangelands, thus exacerbating the existing erosion and making it more difficult to sustain soil productivity.

One of the cited advantages of conservation tillage and no-till, moisture conservation, can be a disadvantage with certain soil conditions. Problems arise with poorly drained soils, where the excessive moisture becomes compounded by reduced tillage. Fortunately, drainage does appear to be a main factor in determining the adoption of conservation tillage and no-till (Crosson 1981). The increased moisture can also pose a problem even on soils with reasonable drainage because moist soils are slower to warm up in the spring. In northern areas the need to delay planting until soils are warmed combined with a short growing season can limit the use of reduced tillage. One solution might be the use of winter cover crops or sod crops that would dry the soil in early spring through evapotranspiration.

Communications also pose some problems. A significant number of nonusers of conservation tillage or no-till have misconceptions about the advantages of reduced tillage (OTA 1982). If these nonusers were aware of the advantages, they might adopt the practice. Another information problem may arise in the area of different management skills required by users of reduced tillage. Fortunately, acquiring these skills is no more costly or more difficult than acquiring those needed for conventional skills. As the benefits of reduced tillage are increasingly spread through written and oral channels and actually observed, the problem of poor communications will be resolved.

For example, the use of nitrogen fertilizer can be less efficient with no-till systems. Under wet conditions, denitrification may be higher with no-till soils than with conventionally tilled soils (Smith and Rice 1983). However, a change in application time from planting time to 4–6 weeks later minimizes the chances of dentrification. Under wet conditions in early spring denitrification activity is maximal at planting time, but can be easily avoided by delayed fertilization.

Nutrient pollution may pose a final problem. Insufficient data exist about differences in nutrient pollution found with conventional and reduced tillage. Although the volume of runoff and sediment is less with reduced tillage, the concentration of soluble nutrients from decomposition of crop residues and wash-off of fertilizers both from crops and surface crop residues may be greater. Wash-off can be severe if rainfall occurs shortly after fertilization. Increased filtration from crop residues may further aggravate the problem. Additional research may resolve the question.

Considering the possible extensive reliance on conservation tillage and no-till, the inadequacy of the data base is surprising. Clearly the extent of reduced tillage must be documented better in terms of acreage, type of land, and the extent of erosion before and after. Much work is needed on cost benefits of erosion control versus the possible costs of increased pollution associated with reduced tillage. The actual extent of nutrient and pesticide pollution arising from reduced tillage needs to be documented. Indeed, the present lack of data may well be the worst problem.

Older Practices for Erosion Control

Conservation tillage and no-till which are relatively new practices, offer substantial help in controlling erosion but do not solve the entire problem. For example, they are not useful on poorly drained soils or in the upper northern states with short growing seasons. In addition these practices would be insufficient on areas subject to higher rates of erosion. Many older practices of erosion control could be used on areas not amenable to reduced tillage, or could be used to supplement reduced tillage on the more fragile lands. About half of the 59 million ha (146 million acres) of cropland with severe erosion is unprotected by practices that control erosion.

Terraces.   Terraces (Fig. 7.5) have been used for centuries to control water erosion. Their first recorded use in U.S. agriculture dates back to 1855. Essentially, terraces are earth embankments, channels, or combinations of both constructed across the slope of the land. Several types exist and are discussed by Troeh et al. (1980). They are of great value in controlling erosion and conserving moisture.

Terraces are the most effective mechanical means of erosion control on slopes planted to continuous row crops. Their efficiency is high, in that they can trap up to 85% of the sediment eroded from a field. On the average erosion is reduced 71% on the approximately 7% of the U.S. cropland with terraces (OTA 1982).

However, only a small part (13.7%) of the cropland considered most fragile (37 tons eroded per hectare per year or 15 tons per acre per year), is terraced. There are several reasons for this. Installation costs can be high, often running about $1000/ha ($400/acre). This cost does not include costs attributed to lost cropland, lost crops during construction, increased costs of working terraced cropland, and maintenance. Although the savings resulting from reduced erosion helps to offset these costs, they are not great enough

FIG. 7.5.   Parallel level bench terraces can be costly to construct but prevent serious erosion of cropland. Cost sharing by the government is used to encourage this practice.
Courtesy U.S. Department of Agriculture/Soil Conservation Service

to make terraces economically attractive. In fact terraces appear to present a net loss to farmers (Mitchell *et al.* 1980).

Other problems also with terraces, soil compaction and losses of topsoil, may occur during construction. In addition, some sites are not suitable for terraces. These include sandy soils, stony soils, shallow soils over bedrock or fine-textured impermeable subsoils, areas with complex slopes, and slopes in excess of 8–12%.

Diversions.   These are simple ditches or channels ridged on the downhill side and constructed at the bottom or top of steep slopes. Their function is to protect cropland from erosion and flooding by intercepting runoff water, which is then slowed and carried away. Diversions are used to protect about 0.7% of the U.S. cropland.

Contour Plowing and Planting.   This conservation practice is more popular than terraces for slopes because of its lower cost. Plowing and planting is done on a line perpendicular to the slope of the land. Contour farming (Fig. 7.6) can reduce soil loss by 50% on slopes of 2–8% and not more than 300 feet (91 m) long. Average rates of erosion on contour-farmed land appear to be around 61% less than on similar, but untreated cropland

FIG. 7.6. In contour plowing, Plowing and planting are done across the slope and follow contour lines. This form of tillage has two purposes: water conservation and reduction of erosion. The former benefit makes this practice useful for the renovation of rangeland.
Courtesy U.S. Department of Agriculture/Soil Conservation Service

(OTA 1982). Troeh *et al.* (1980) cover the details of contour-farming practices.

A variation of contour plowing and planting is contour stripcropping (Fig. 7.7). With this practice, contour plowing and planting is used, but continuous row crops are replaced by strips of row crops alternating with strips of forage crops. Row crop strips must be sized to minimize runoff and soil erosion, and the forage strips must be wide enough to slow and filter the runoff. Erosion reduction is about 50% greater than with conventional contour planting. The practice is covered by Troeh *et al.* (1980). As with any contour practice, it is not suitable for steep slopes or slopes with complex, variable slope gradients. Maximal effectiveness in erosion control is obtained by combining terraces and contour stripcropping.

A logical approach is to alternate a perennial legume strip with a row crop. This system eliminates the cost of annual seeding, contributes nitrogen, provides year-round erosion protection, and still provides renewable animal feed. Recently, Palada *et al.* (1983) have examined this concept, which they refer to as sod strip intercropping. Their management system involved either corn or soybeans intercropped with strips of red clover, white clover, or alfalfa. Soybean yields were unaffected with both clovers, but corn yields showed significant reductions.

Certain tillage widths and row patterns were found to minimize the reduction in corn yield. The lowest yield reduction, compared with monoculture corn, was found with corn tillage widths of 0.75 m. Single rows were spaced at 1-m distances, thus the area in clover sod was 25%. If double rows were desired, the tillage width was 1.5 m; double row spacing, 2 m; and area in sod, 25%. The yield reduction was 16% compared with monoculture. Corn density was 40,000 plants/ha (16,194/acre).

Various corn and alfalfa tillage widths and patterns were also investigated by Palada *et al.* (1983). The grain and hay requirements would determine the adjustment to tillage width and fraction of alfalfa sod. Corn yields were reduced compared with a monoculture system, even when supplemented with nitrogen fertilizer up to 240 kg/ha (214 lb/acre). The relative yield at this rate was 64%.

While these yield reductions are unacceptable to cash-grain farmers, they might be reasonable for farmers utilizing grain and hay for animal feed. This management system would be especially attractive for slopes currently in complete forage, since it would allow some row crop production but offer more erosion control than is possible with only row crops. More research is needed to determine the best spacing and percentage of sod strips needed to maximize yields and minimize erosion.

FIG. 7.7.   Contour stripcropping is more effective for erosion reduction than contour planting, but yields of the main crop are less. However, the alternating forage crops compensate in part.
Courtesy U.S. Department of Agriculture/Soil Conservation Service

Cover Crops and Crop Rotations. These practices minimize both water and wind erosion. Both help to keep a continuous cover on fields, thus reducing susceptibility to erosion. Cover crops, often cereal grains such as oats, rye, or winter wheat, are planted between regular cropping periods. With crop rotations forage crops are rotated with nonforage crops. These practices reduce erosion over continuous cropping by 25–30%. If correctly chosen, the forage crops can also supply nitrogen for the crops that follow, as was discussed in Chapter 3.

Rotations do tie up part of the cropland, but the forage value provides some compensation. Some farmers may prefer a combined system in which some food crops are harvested from the land currently planted to the forage portion of the rotation cycle. This established practice of combining crops and forages is currently receiving renewed attention (Palada *et al.* 1983). Two approaches exist at present: One involves overseeding of a legume with either corn or soybeans, and the other involves vegetables combined with a legume in a living mulch system.

Palada *et al.* (1983) found that overseeding of legumes (various clovers, hairy vetch) roughly 5 weeks after row crop seeding and near the first, rather than second cultivation, resulted in better germination, higher seedling emergence, good weed control, and improved fall and winter ground cover. Yields of corn and soybeans were not reduced. Seeding densities were 40,000 and 60,000 plants/ha (16,194 and 24,291/acre) for corn and 400,000 plants/ha (161,943/acre) for soybeans. Ground cover by the legume was considerably better with corn than with soybeans because of the low light intensity under the soybean canopy.

In the spring the legume ground cover can be plowed under to supply nitrogen for the following crop, or allowed to continue as a forage legume rotation. This flexibility allows the farmer some advantage in cash returns, as he may change the pattern to meet his forage needs or respond to market values of crops and forages.

Vegetable–living mulch systems also offer an alternative way to harvest both crops and forages. Palada *et al.* (1983) have examined the strip planting of corn and tomatoes in grass and clover sods. They compared mulched tilled strips to mulched no-tillage strips. In the latter, sod suppression was achieved by either alfalfa or black plastic mulch, not by herbicides. Tomatoes showed the highest yields in the alfalfa mulched strips in both the tilled and no-till strips, as compared to black plastic and clean cultivation treatments. Corn yields, however, were best in the clean cultivation strips.

Jurchak (1984) also examined vegetable–living mulch systems. He tested legumes and grasses, looking for the annual or perennial that offered the most soil improvement and erosion protection and the least competition to the vegetable row crop. Annual ryegrass was recommended as the best choice. This grass reached maximal development after the crop was harvested, provided a winter cover, and increased soil organic matter when it was plowed under in the spring.

This approach has been used with several vegetable crops: beans, broccoli, cabbage, cauliflower, peppers, potatoes, sweet corn, and tomatoes. Some 162 ha (400 acres) are managed under this system with tomatoes being the biggest crop. Costs are shared between the growers and Agricultural Stabilization and Conservation Service in Lackawanna County, Pennsylvania. Long-term studies of the system are underway.

All of these systems appear promising, but more research is needed to establish the best species of legumes for optimal nitrogen management. Other crops should also be tested. The contribution of the combined system in terms of nitrogen savings and erosion control should be clearly documented.

Grass Waterways.   Strips of land covered with grass can be utilized as paths for transporting surface runoff from fields at nonerosive velocities. Grass waterways are very common, used alone or in combinations with terraces or diversions. Maintenance can be difficult, as herbicides in the runoff can destroy the grass.

Other Practices.   Several other practices help to reduce erosion; these include reduction of field lengths, stripcropping, windbreaks, shelterbelts and mulches. Wind erosion can be reduced by shortening field lengths along the direction of the prevailing wind. Alternation of strips of crops susceptible and resistant to wind erosion at right angles to the prevailing winds is termed stripcropping. Trees can be planted as windbreaks and shelterbelts to lower wind speeds. Mulches can be used to cover and protect the soil against wind erosion. These practices are detailed by Troeh *et al.* (1980).

Soil Compaction

As farm machinery has become increasingly heavier with the increase in size and efficiency of U.S. farms, concern about compaction of soil on cropland and rangeland has grown. The effects of compaction on cropland and rangeland differ somewhat.

On cropland compaction creates a traffic pan, an area of densely compacted soil underlying the depth to which the soil is tilled. Since compaction collapses soil macropores, this interferes with movement of soil water. In turn this can lead to standing water and increased erosion of soil by surface runoff. Several subsequent problems result: Soils remain colder longer, causing delayed planting and slower seed germination; crop residue decomposes slower and losses of soil nitrogen from denitrification tend to increase;

root development can be hindered, thus water and nutrient availability becomes a problem.

Usually such conditions lead to reduced crop yields (OTA 1982); for example, corn yields may be up to 50% less on compacted clay soils than on similar uncompacted soils. The extent of crop yield decreases can also be shown by measuring yields after removing the traffic pan by subsoiling (deeper tillage). Under these conditions yield increases of up to 83% have been noted for cotton and corn. In another study, corn yields increased by 247 bu/ha (100 bu/acre) after subsoiling.

Such effects are not noticed by farmers all that much, since they are usually masked by increased use of irrigation and fertilization. However, increased costs for these inputs is directly attributed to soil compaction. In some instances compaction may mask soil deterioration resulting from decreasing organic matter content. Such losses lessen the aggregation properties associated with organic matter, but compaction can lead to slight increases in soil aggregate size and stability—a temporary, beneficial effect.

Other beneficial effects have been found to be associated with moderate compaction (OTA 1982) of dry soils. Dry soils can warm earlier in the spring and the traffic pan helps to retain surface water for seed germination. Yields of soybeans are higher on moderately compacted soils in dry years than on soils without compaction. Some crops such as corn mature earlier and have a lower grain moisture content on moderately compacted soils.

Still, on the whole, compaction is not desirable, and a number of ways can be used to minimize or correct it. If machinery passes on a field are less frequent, compaction is reduced. Minimal compaction is found with conservation tillage and no-till, tillage systems already noted to reduce soil erosion and energy consumption. Existing compaction can be corrected by deep tillage, called subsoiling. Deeper tillage beyond 7–8 in. (17.8–20.3 cm) breaks the traffic pan and improves aeration and percolation (Cassel 1979). Large flotation tires also minimize compaction.

On rangeland animal traffic dominates, leading to compression of surface soil. This compaction effect is termed shingling. Infiltration of water is greatly reduced, leading to little or no plant cover. Erosion susceptibility becomes greater, either by wind or water. Off-road vehicles can lead to compacted strips devoid of plants and highly susceptible to gulley erosion. Reduction of plant cover means lower rangeland productivity. A further complication arises on rangeland devoid of plant cover. Raindrops impacting on unprotected soil produces a thin crust, a process called soil capping. This process further exacerbates the shingling effect.

Correction of these problems is possible. First, the number of cattle must be reduced to match the carrying capacity of the rangeland. Depending on the severity of the compaction, this might be sufficient. If not, reseeding may be necessary. In very severe cases, tillage may be needed to break up the shingling prior to reseeding. The expense of the latter tends to limit its feasibility.

Much research is needed on the degree to which compaction limits the productivity of cropland, and even more so for rangeland. Machinery needs to be designed for minimal compaction; wide-span equipment is one such option (OTA 1982). The economic data on losses of productivity versus correction of compaction are quite limited.

Cropping Systems to Sustain Productivity

The trend in American agriculture has been toward monoculture, with a subsequent loss in crop variety and crop rotation. This trend has not necessarily been beneficial in terms of inherent land productivity because it leads to increased insect and disease problems and increased soil erosion, especially with row crops. One alternative to monoculture, multiple cropping, is already in practice and will probably increase in the future. Crop rotations are returning, and may assume greater importance in the future.

Multiple Cropping

The productivity of cropland is maximized by multiple cropping, since the practice makes more efficient use of space and time than monocropping. This explains why the overall yield of the combined crops is greater, even though some may yield slightly less than they would in monoculture. Careful thought must be given to the selection of the crops so that they are complementary in terms of needs for sunlight, water, and nutrients. Three basic variations of multiple cropping are practiced: sequential cropping, and relay and simultaneous intercropping.

The easiest way to insure complementarity in a multiple-cropping system is to plant crops in sequence. The most widely practiced form of sequential cropping is double cropping; a common system is barley or wheat followed by soybeans.

Adoption of reduced tillage systems has made double cropping feasible in areas with marginal growing seasons and water availability because of the time savings and moisture conservation possible with reduced tillage. Areas with increased double cropping resulting from the use of reduced tillage and shorter-season cultivars of soybeans include Delaware, Maryland, and the southern part of the Corn Belt. Double cropping is practiced extensively in the southeastern U.S., which has long growing seasons (ASA 1976; OTA 1982). Double cropping also is common in northwestern California and western Oregon and Washington where mild winters make it possible. Crop combinations here include oats and red clover, forage crops with young orchard trees, and various combinations of vegetables.

The other two approaches to multiple cropping are relay and simultaneous intercropping. Relay intercropping involves planting a second crop before harvesting the first crop but after the first crop has completed the major part of its development. With simultaneous intercropping the crops are planted simultaneously but usually mature at different times. An example of relay intercropping is no-till wheat (first crop) and soybeans. Corn, squash, and beans can be cropped simultaneously. Horwith (1985) discusses the possible role of intercropping in future agriculture.

If the selection and management of multiple-cropping systems is optimal, a number of benefits accrue, yet cropland productivity is sustained. Wind and water erosion is reduced with double cropping and relay intercropping because the soil is covered by crops for a longer period of time than with monoculture. Some reduction of erosion results from simultaneous intercropping because a greater part of the soil surface is covered.

Total marketable yields and income are often greater with multiple cropping than with monoculture. This is true for double cropping of small grains with short-season cultivars of soybeans, as opposed to full-season soybeans. Increased forage production is also possible with double cropping of small grains and corn or sorghum. Both may be grown for silage, or one for grain and the other silage. Even greater yields are possible with triple cropping of barley, corn, and soybeans or snapbeans. The latter requires a long growing season, such as exists in Georgia (ASA 1976). Obviously fertilization and pest control require careful control to assure high yields and sustainability of the land.

To achieve optimal yields and income, the crops in a multiple-cropping system must have sufficient differences in their developmental needs so that competition for sun, water or nutrients is either a minimal or negligible constraint. Under these conditions multiple cropping can increase cropland productivity through optimal utilization of the available time and space (OTA 1982). Thus, multiple cropping can lead to increased food supplies without decreasing the sustainability of the soil. In addition, such increased production from multiple cropping will reduce the pressure for expansion of crop production into areas having a high susceptibility to erosion. Sloping cropland with high rates of erosion because of row crops may even be returned to hay and pasture crops with subsequent decreases in erosion if multiple cropping becomes prevalent.

Other advantages may exist, depending upon the crop choice and form of multiple cropping. Diversity of crops rather than monoculture, can lead to reduced susceptibility to insects and diseases, and increased crop cover and competition, such that weeds are decreased. If one crop is a legume, fertilizer requirements can be reduced. If some crops are deep rooted, a number of advantages incur. Nutrients are used more efficiently, as those leached by or present beneath the shorter root zone of one crop can be utilized by the deeper roots of the other crop. Deep penetration of roots helps to improve aeration and water infiltration, and lessens the possibility of hardpan formation. More crops may increase surface cover, thus reducing evaporation from the soil. Diversity may promote increased beneficial microbial activity in the soil. Finally, the organic matter content of the soil is more easily maintained because of increased crop residues.

Multiple cropping may have some disadvantages depending upon circumstances. Poor choices of crops can result in excessive competition for water, light, and nutrients, leading to lower yields. Improper monitoring of soil nutrients can also lead to reduced yields, as can neglecting to apply sufficient fertilizers or to use rotations with legumes. Such situations can lead to harm of inherent land productivity. The presence of mixed crops also may increase the difficulty of such mechanized operations as tillage, planting, and harvesting. For example, operations on one crop must not harm the other. This can be minimized by the correct choice of crops or by modification of machinery (ASA 1976).

The disadvantages of multiple cropping are not great enough to offset the advantages, as can be seen by the increasing prevalence of double cropping in the United States (OTA 1982), although little data exist to substantiate the actual extent or annual increase of multiple cropping. In light of its tangible benefits, more research is needed to learn about the complicated agroecosystems involved in multiple cropping. Such data are needed to further minimize its disadvantages and to encourage widespread adoption of multiple cropping by U.S. farmers.

Rotations, Nitrogen Fixation, and Forages

The major concern with the sustainability of soil nutrients is not lack of nutrients (fertilizers are widely available) but the energy consumption associated with fertilization, as detailed in Chapter 3. In that chapter, possibilities leading to savings of energy were discussed. The use of organic amendments, discussed in this chapter, serves not only to maintain organic matter but to also supply nutrients. Unhappily they are not the entire answer, as the most available ones (e.g., manures and crop residues) are almost used to capacity already and increased usage of the others is not highly feasible. A further problem is competition for their use as energy sources and feeds.

One answer to sustaining nutrients, but keeping energy consumption down, is to return to some older farming practices. These fell into disfavor with the advent of inexpensive fertilizers and the adoption of monoculture, because of the latter's efficiency and profitability. The slighted practices included crop rotations and cover crops. The rotations involved food crops and forages. Nitrogen needs were satisfied by working legumes into the rotation, since they are capable of nitrogen fixation (see Table 3.3). It should also be remembered that rotations and cover crops can help reduce soil erosion.

A case in point would be continuous cropping of corn, which gradually replaced corn in rotation with other crops in the years following World War II. Such rotations involving corn were common back in the 1930s. However, erosion rates with corn alone are 25–30% higher than with corn in rotation. Corn is not a soil-conserving crop, but if the rotation includes a soil-conserving crop, then the soil is protected while this crop is growing. Often a carryover effect is noted, such that the erosion normally found with corn is slighly less.

Continuous cropping of corn also leaves the soil exposed in the winter, but if the rotation utilizes a hardy forage crop, the soil is protected with a cover crop during the winter. Such a crop could be plowed under as a green manure in the spring, or allowed to stand as a temporary pasture or used for hay production. Green manures are beneficial because they increase organic matter content and improve the structure and permeability of the soil.

Legume crops are popular in corn rotations because they reduce nitrogen. Some legumes, such as soybeans, fix about one-third of their nitrogen, but they leave essentially no nitrogen in the soil for succeeding crops, since most of it is lost with the harvested portions. Nevertheless, corn–soybean rotations are possible, since nitrogen needs are less and both crops are in high demand.

However, forage legumes leave substantial amounts of nitrogen in the soil for crops that follow. They also supply about 80% of their nitrogen needs via nitrogen fixation. Nitrogen is returned to the soil from the root system and unharvested regrowth. One of the better forage legumes for rotations is alfalfa. Rotations in the Midwest might include oats/3 years of alfalfa/corn or wheat/soybeans/corn/soybeans (OTA 1982). Other examples (Heichel 1978) include 2 years of corn/soybeans; 2 years of corn/oats/2 years of alfalfa; 2 years of corn/alfalfa; 3 years of corn/3 years of soybeans/wheat/3 years of alfalfa; and corn/soybeans/vetch.

Rotations are critized on the basis that they are not as economically efficient as corn. Still, rotations are becoming increasingly cost effective and commonly acceptable as a viable alternative to continuous corn for several reasons. The demand for forage crops helps, as does the reduced energy needs in the short run. For the long run, the value of rotations lies in erosion reduction. More research is needed to establish a better data basis for the evaluation of various rotations.

Organic Agriculture and Soil Maintenance

Although organic agriculture was covered in Chapter 2, it will be briefly mentioned here, since it has many advantages for sustaining inherent land productivity (OTA 1982). Adequate nitrogen is supplied from manures, crop residues, and crop rotations utilizing cover crops and legumes. This system not only cuts energy needs but reduces erosion by one-third or more.

The forages from the rotations are used most economically on organic farms by direct feeding to livestock. Most of the successful organic farms are mixed crop and livestock operations. Livestock also supply manures directly on the farm, thus minimizing costs of transport. Pollution problems are also minimal because nitrogen runoff is less, as is pesticide runoff.

Additional research is needed to investigate organic agriculture. Many of its practices are of value not only to organic farmers, but to any farmer who wishes to practice sustainable agriculture. It appears that agricultural management systems utilizing the basic practices of organic farming and small amounts of manufactured fertilizers may have the potential to become the basis for the sustainability of inherent land productivity (OTA 1982).

Maintenance of Range Productivity

At present insufficient data are available to determine the overall productivity of rangeland. One problem that complicates the determination of rangeland status is the highly variable weather associated with much of the rangeland. Such variability can produce a threefold variation in rangeland plant production from one year to the next. Future monitoring of rangeland on a regular basis, including several periods of drought, is needed for proper assessment (OTA 1982).

Certainly some part of the total U.S. rangeland has deteriorated. A number of rangeland management systems are available for improvement of deteriorated rangeland and for maintenance of inherent rangeland productivity. These include the adjustment of livestock numbers to the land's carrying capacity, the manipulation of rangeland vegetation to increase the vigor and abundance of desirable grazing plants, deferred and rotational grazing, and the control of undesirable plants and animals. A good discussion of these current rangeland technologies is available (OTA 1982).

Some newer systems are also evolving. One is an integrated brush management system (IBMS), which utilizes a combination of brush control methods, rather than just one control method. Any one method of brush control often has drawbacks: for example, physical removal can become expensive because of labor and energy costs, and heavy dependence on herbicides can be costly and damaging to the environment. Essentially IBMS uses a mix of both old and new forms of brush control to capitalize on the individual strengths of each and minimize the overall disadvantages. A good discussion of IBMS is available (Scifres 1980).

Another innovation discussed by Scifres (1980) is short-duration grazing systems. In this approach larger numbers of animals are grazed for a shorter time on a given area than is common with conventional grazing. A number of areas in rotation are subjected to short-duration grazing, thus providing time for regeneration. The claim is that short-duration grazing increases carrying capacity without decreasing rangeland productivity. One variation of the short-duration grazing system is the Savory grazing method (OTA 1982). Rotated pastures are arranged in a "cartwheel" design around a central hub, which contains the watering and handling facilities.

All of these innovative systems need extensive investigation to determine their potentials, as well as their constraints. An excellent examination of these approaches and other range management systems in the context of resources, policy, and socioeconomics is available (NRC/NAS 1984).

References

ANON.   1984.   Idaho: reducing runoff on potato land. Amer. Veg. Grower *32* (1) 20.

ASA.   1976.   Multiple Cropping. Special Publication No. 27. Amer. Soc. Agron., Madison, WI.

BARBARIKA, A., COLACICCO, D. and BELLOWS, W. J.   1980.   The value and use of organic wastes. Maryland Agri-Economics, Maryland Coop. Ext. Service.

BROWN, L. R.   1981.   Building a Sustainable Society. W. W. Norton Co., New York.

BUNT, A. C.   1976.   Modern Potting Composts. Penn. State Univ. Press, University Park.

BUNTLEY, G. T. and BELL, F. F.   1976.   Yield estimates for the major crops grown on soils of west Tennessee. Tenn. Agric. Exp. St. Bull. *561*.

BURGE, W. D., COLACICCO, D. and CRAMER, W. N.   1981.   Criteria for achieving pathogen destruction during composting. J. Water Pollut. Control Fed. *53*, 1683–1690.

CASSEL, D. K.   1979.   Subsoiling. Crops and Soils Magazine (October), 7–10.

CAST.   1982.   Soil Erosion: Its Agricultural Environmental, and Socio-economic Implications. Rept. No. 92. Council for Agric. Sci. and Technol., Dept. of Agronomy, Iowa State Univ., Ames.

CROSSON, P.   1981.   Conservation and Conventional Tillage: A Comparative Assessment. Soil Conservation Soc. of America, Ankeny, Iowa.

EPSTEIN, E.   1975.   Effect of sewage sludge on some soil physical properties. J. Environ. Qual. *4*, 139–142.

EPSTEIN, E., TAYLOR, J. M. and CHANEY, R. L.   1976.   Effects of sewage sludge and sludge compost applied to soil on some soil physical and chemical properties. J. Environ. Qual. *5*, 422–426.

FRYE, W. W., HERBEK, J. H. and BLEVINS, R. L.   1983.   Legume cover crops in production of no-tillage corn. *In* Environmentally Sound Agriculture. W. Lockeretz (Editor). Praeger Publishers, New York.

GILBERTSON, C. B., NORSTADF, F. A., MATHERS, A. C., HOLT, R. F., BARNETT, A. P., McCALLA, T. M., ONSTAD, C. A. and YOUNG, R. A.   1979.   Animal Waste Utilization on

Cropland and Pastureland: A Manual for Evaluating Agronomic and Environmental Effects. Science and Education Admin., U.S. Dept. Agric., Washington, DC.

GREENHOUSE, S.   1984.   Quiet revolution on the farm. New York Times (November 5), D1, 15.

HAGEN, L. L. and DYKE, P. T.   1980.   Yield–soil loss relationship. *In* Proc. of Workshop on Influence of Soil Erosion on Soil Productivity. USDA-SEA-AR, Washington, DC.

HEICHEL, G. H.   1978.   Stabilizing energy needs: Role of forages, rotations and nitrogen fixation. J. of Soil Water Conserv. *33* (6) 279–282.

HOITINK, H. A. J. and POOLE, H. A.   1980.   Factors affecting quality of composts for utilization in container media. *In* Proc. of Symposium on Using Municipal and Agricultural Waste for the Production of Horticultural Crops. A special insert in HortScience *15*(2).

HORNICK, S. B.   1982.   Use of organic waste materials in the revegetation of marginal lands. BioCycle J. Waste Recycling *23* (4) 42–43.

HORNICK, S. B., MURRAY, J. J., CHANEY, R. L., SIKORA, L. J., PARR, J. F., BURGE, W. D., WILLSON, G. B. and TESTER, C. F.   1979.   Use of Sewage Sludge Compost for Soil Improvement and Plant Growth. Science and Education Administration, Agricultural Reviews and Manuals, Northeastern Region Series 6, U.S. Dept. Agric., Beltsville, MD.

HORWITH, B.   1985.   A role for intercropping in modern agriculture. BioScience *35*(5) 286–291.

JURCHAK, T.   1984.   Improve your soil with living mulch. Amer. Veg. Grower *32,* 40–41.

KHALEEL, R., REDDY, K. R. and OVERCASH, M. R.   1981.   Changes in soil physical properties due to organic waste applications: A review. J. Environ. Qual. *10,* 133–141.

KIRKBY, M. J. and MORGAN, R. P. C.   1980.   Soil Erosion. John Wiley & Sons, New York.

LARSON, W. E.   1981.   Protecting the soil resource base. J. Soil Water Conserv. *36* (1) 13–16.

LARSON, W. E., HOLT, R. F. and CARLSON, C. W.   1978.   Residues for soil conservation. *In* Crop Residue Management Systems.   W. R. Oschwald (Editor).   Amer. Soc. Agron., Madison, WI.

LARSON, W. E., PIERCE, F. J. and DOWDY, R. H.   1983.   The threat of soil erosion to long-term crop production. Science *219,* 458–465.

LAUER, D. A., BOULDIN, D. R. and KLAUSNER, S. D.   1976.   Ammonia volatilization from dairy manure spread on the soil surface. J. Environ. Qual. *5,* 134–141.

LUMSDEN, R. D., LEWIS, J. A., WERNER, R. E. and MILLNER, P. D.   1981.   Effect of sludge compost on selected soilborne diseases. Phytopathology *71,* 238.

McCORMACK, D. E. and LARSON, W. E.   1980.   A values dilemma—Standards for soil quality tomorrow. Paper presented at Soil Conserv. Soc. Amer., Dearborn, MI.

MERKEL, J. A.   1981.   Managing Livestock Wastes. AVI Publishing Co., Westport, CT.

MILLNER, P. D.   1982.   Thermophilic and thermotolerant actinomycetes in sewage sludge compost. Dev. Ind. Microbiol. *23,* 61–78.

MILLNER, P. D., BASSETT, D. A. and MARSH, P. B.   1980.   Dispersal of *Aspergillus fumigatus* from sewage sludge compost piles subjected to mechanical agitation in open air. Appl. Environ. Microbiol. *39,* 1000–1009.

MITCHELL, J. K., BRANCH, J. C. and SWANSON, E. R.   1980.   Costs and benefits of terraces for erosion control. J. Soil and Water Conserv. *35* (5) 233–236.

NRC/NAS.   1984.   Developing Strategies for Rangeland Management. National Research Council/National Academy of Sciences Report. Westview Press, Boulder, CO.

OTA.   1982.   Impacts of Technology on U.S.—Cropland and Rangeland Productivity. Office of Technology Assessment, Washington, DC.

OWEN, E.   1980.   Agricultural wastes as feedstuffs. *In* Handbook of Organic Waste Converson. M. W. M. Bewick (Editor). Van Nostrand Reinhold Co., New York.

PALADA, M. C., GANSER, S., HOFSTETTER, R., VOLAK, B. and CULIK, M. 1983.   Association of interseeded legume cover crops and annual row crops in year-round cropping systems. *In* Environmentally Sound Agriculture. W. Lockeretz (Editor). Praeger Publishers, New York.

PARR, J. F. 1982. Composted sewage sludge, a potential resource for small farms. *In* Proc. of a Special Symposium, Research for Small Farms. USDA-ARS Misc. Pub. No. 1422.

PARR, J. F. and PAPENDICK, R. I. 1983. Strategies for improving soil productivity in developing countries with organic wastes. *In* Environmentally Sound Agriculture. W. Lockeretz (Editor). Praeger Publishers, New York.

PARR, J. F., MARSH, P. B. and KLA, J. M. 1983. Land Treatment of Hazardous Wastes. Noyes Data Corp., Park Ridge, NJ.

PHILLIPS, R. E., BLEVINS, R. L., THOMAS, G. W., FRYE, W. W. and PHILLIPS, S. H. 1980. No-tillage agriculture. Science *208,* 1108–1113.

PIMENTEL, D., TERHUNE, E. C., DYSON-HUDSON, R., ROCHERCAU, S., SAMIS, R., SMITH, E. A., DENIMAN, D., REIFSCHNEIDER, D., and SHEPARD, M. 1976. Land degradation: Effects on food and energy resources. Science *194,* 149–155.

POINCELOT, R. P. 1975. The biochemistry and methodology of composting. Conn. Agric. Exp. Stn. Bull. *754.*

POINCELOT, R. P. 1980. Horticulture: Principles and Practical Applications. Prentice-Hall, Englewood Cliffs, NJ.

REDDY, K. R., KHALEEL, R. and OVERCASH, M. R. 1981. Behavior and transport of microbial pathogens and indicator organisms in soils treated with organic wastes. J. Environ. Qual. *10,* 255–266.

ROSENBERRY, P., KNUTSEN, R. and HARMON, L. 1980. Predicting the effects of soil depletion from erosion. J. Soil and Water Conserv. *35* (3) 131–134.

SANDERSON, K. C. 1980. Use of sewage-refuse compost in the production of ornamental plants. *In* Proc. of Symposium on Using Municipal and Agricultural Waste for the Production of Horticultural Crops. A special insert in HortScience *15*(2).

SAXTON, K. E., McCOOL, D. K. and PAPENDICK, R. I. 1981. Slot mulch for runoff and erosion control. J. Soil Water Conserv. *36* (1) 44–47.

SCIFRES, C. J. 1980. Emerging Innovative Technologies for Rangeland. OTA Background Paper. National Technical Information Service, U.S. Dept. of Commerce, Springfield, VA.

SIMS, J. T. and BOSWELL, F. C. 1980. The influence of organic wastes and inorganic nitrogen sources on soil nitrogen, yield, and elemental composition of corn. J. Environ. Qual. *9,* 512–518.

SMITH, M. S. and RICE, C. W. 1983. Soil biology and biochemical nitrogen transformations in no-tilled soils. *In* Environmentally Sound Agriculture. W. Lockeretz (Editor). Praeger Publishers, New York.

STERRETT, S. B., REYNOLDS, C. W., SCHALES, F. D., CHANEY, R. L. and DOUGLASS, L. W. 1983. Transplant quality and heavy metal accumulation of tomato, muskmelon, and cabbage grown in media containing sewage sludge compost. J. Amer. Soc. Hort. Sci. *108* (1) 36–41.

TESTER, C. F. and PARR, J. F. 1983. Intensive vegetable production using compost. BioCycle J. Waste Recycling *24* (1) 34–36.

TRIPLETT, JR., G. B. and van DOREN, JR., D. M. 1977. Agriculture without tillage. Sci. Amer. *236* (1), 28–33.

TROEH, F. R., HOBBS, J. A. and DONAHUE, R. L. 1980. Soil and Water Conservation for Productivity and Environmental Protection. Prentice-Hall, Englewood Cliffs, NJ.

USDA. 1978. Improving Soils With Organic Wastes. Office of the Secretary of Agriculture, Washington, DC.

USDA. 1980. Soil and Water Resources Conservation Act., Summary of Appraisal. Parts I and II and Program Report. U.S. Dept. of Agric., Washington, DC.

USDA. 1981. Soil and Water Resources Conservation Act: Program Report and Environmental Impact Statement. Revised draft. U.S. Dept of Agric., Washington, DC.

USDA. 1982. Agricultural Statistics 1981. U.S. Dept. of Agric., Washington, DC.

USDA. 1985. Research Progress in 1984. U.S. Dept. of Agric., Washington, DC.

USDA/CEQ. 1980. Soil Degradation: Effects on Agricultural Productivity. Interim Report No. 4. U.S. Dept. of Agric. and the Council on Environ. Quality, Washington, DC.

USDA/EPA. 1980. Manual for Composting Sewage Sludge by the Beltsville Aerated-Pile Method. EPA-600/98-80-022. A special report available through the National Technical Information Service, Springfield, VA.

WALKER, J. M. 1980. Government regulations on the use of municipal organic materials on agricultural lands. *In* Proc. of Symposium on Using Municipal and Agricultural Waste for the Production of Horticultural Crops. A special insert in HortScience *15*(2).

WILLIAMS, J. R., ALLMARAS, R. R., RENARD, K. G., LYLES, L., MOLDENHAUER, W. C., LANG-DALE, G. W., MEYER, L. D., RAWLS, W. J., DARBY, G. and DANIELS, R. 1980. Soil erosion effects on soil productivity: A research perspective. J. Soil and Water Conserv. *35* (2), 82–90.

WILLSON, G. B., PARR, J. F., TAYLOR, J. M. and SIKORA, L. J. 1982. Land treatment of industrial wastes: Principles and practices. Parts I and II. Biocycle J. Waste Recycling *23*(1), 37–42 and *23* (2), 59–61.

WISCHMEIER, W. H. and SMITH, D. D. 1978. Predicting Rainfall Erosion Losses—A Guide to Conservation Planning. Agric. Handbook 537. U.S. Dept. Agric., Washington, DC.

8

Sustaining Resources: Water

Shortages of water—the most limited of all agricultural resources—are more likely to cause agricultural problems in the near future than are soil erosion and environmental pollution. Competition for water occurs among the agricultural, industrial, public, and energy-producing sectors of the United States. Further restraints upon water supplies are also imposed by the need to maintain aquatic systems and to preserve the environment. As such the sustainability of agricultural water resources is the most pressing problem in U.S. agriculture.

Usage and Availability

Water withdrawal in the United States, exclusive of hydroelectric use, is estimated at 1600 billion liters/day (420 billion gal./day). Of this, about 600 billion liters/day (157 billion gal./day), or 38%, is used in agriculture; 99% of this is used for irrigation and about 1% for livestock. About 60% of irrigation water is derived from surface water and 40% from groundwater (Pimentel et al. 1982; USDA 1981).

However, a closer examination of these figures points out the serious nature of water use in agriculture. Of the total water withdrawn, about three-fourths is returned to lakes and rivers; the remainder, 400 billion liters/day (106 billion gal./day), is consumed or lost. Irrigation accounts for 83% of the water consumed and lost. This extensive water usage is confined to the roughly 12% of the nation's cropland that produces about 27% of the total dollar value of U.S. crops (USDA 1981; OTA 1982; Pimentel et al. 1982).

Most of the irrigated cropland is concentrated in 17 western states. Because these croplands are arid, large amounts of water are needed. Furthermore, high evapotranspiration occurs in these regions, further increasing water usage. These states depend primarily upon groundwater withdrawal, with some states supplying up to 94% of their irrigation needs from groundwater.

Water availability presents a problem in view of current use. According to the U.S. Water Resources Council (USWRC 1979), the maximal and minimal seasonal availability of stream flow water ranges from 2600 to 1350 billion liters/day (687 to 357 billion gal./day). Surface water capacity is unlikely to be increased by construction of additional reservoirs, because of economic and site feasibility restrictions (Pimentel *et al.* 1982). Average renewal times for rivers and freshwater lakes are 12–20 days and 10–100 years, respectively. Estimates place the amount of stored groundwater as 10-fold greater than surface water. Unlike the rapid replenishment of surface water through precipitation, the annual replenishment rate for groundwater averages about 230 billion liters/day (61 billion gal./day), or less than 1% of the total. It takes about 300 years to replace all the groundwater (CEQ 1980) to 1000 m, and 4600 years to replace the groundwater at depths of 1000–2000 m.

Groundwater Depletion

Unfortunately, the withdrawal of groundwater for agricultural use exceeds the recharge rate. Groundwater is essentially being mined. Withdrawal exceeds replenishment on the national scale by about 25% and by as much as 77% in the Texas–Gulf area (Pimentel *et al.* 1982). The U.S. annual withdrawal of groundwater is estimated to be 29.1 trillion liters (7.7 trillion gal.).

Under these conditions water deficits are already found in certain agricultural regions highly dependent upon irrigation. Currently, overwithdrawals are occurring in the western United States, especially in areas along the Colorado River, the San Joaquin Valley, and in the Great Plains (Pimentel *et al.* 1982). Specific states experiencing serious overwithdrawal problems include Arizona, California, Colorado, Hawaii, Kansas, New Mexico, and Texas. Even the eastern United States, blessed with more water, has been hit with spot shortages of water in limited areas (Sheer 1980).

One area of particular concern is the Great Plains, which overlies the Ogallala aquifer. This aquifer is exceptionally large in extent and volume, and especially important to Great Plains states. The states utilizing this aquifer include Colorado, Kansas, Nebraska, New Mexico, Oklahoma, South Dakota, Texas, and Wyoming. This groundwater resource is especially important to the region's agricultural economy. Without this aquifer, the $30 billion annual agricultural economy would be jeopardized. Water from this source is used on some 4.9 million ha (12 million acres) of cropland, and also helps to support nearly 500,000 people. This acreage represents roughly one-fifth of the irrigated U.S. farmland. Almost one-half of the U.S. food supply of beef is fattened in this same area (Pye *et al.* 1983).

Water withdrawals from the Ogallala aquifer are considerable. Annual removal is estimated at 5–8 million acre-ft, which exceeds the entire annual flow of the Colorado River. The water table has fallen considerably in this area. If current usage continues, the aquifer is likely to be depleted in 40 years. Three states are expected to be hit hardest: Texas, New Mexico, and Oklahoma (Pye *et al.* 1983; Miller 1982; OTA 1982). Already, Gaines County in Texas has lost its groundwater resource.

The seriousness of the problem is evident from the following data (OTA 1982). The Ogallala aquifer now varies in thickness from 0.3 to 360 m (1 to 1200 ft). Presently, 46% of the aquifer has less than 30 m (100 ft) of water-filled sediment. Groundwater pumping has resulted in noticeable drops in the water table. One-seventh of the Ogallala area's water table has dropped 3–7.5 m (10–25 ft), and the drop has been as high as 30–45 m (100–150 ft) in 2% of the area.

In response to these developments, a High Plains Study Council, funded through Congress, commenced a comprehensive resource and economic development study in 1977. Proposals for lessening withdrawals from the Ogallala aquifer include conservation, more efficient irrigation, and transfer of surface water from adjacent regions. The latter is complicated by lack of public support and interstate cooperation, wide differences in state water control laws, and prohibitive costs. The most likely outcome is a gradual conversion to dryland farming of the most affected areas. Dryland wheat farming and cattle grazing were once common here, until widespread irrigation took place after World War II (Pye *et al.* 1983).

Future Demand

A number of factors, not all agricultural, will undoubtedly increase the demand upon U.S. water supplies. Modest increases will come from industry and the general public. Agricultural needs for irrigation water are projected to increase by 17% by the year 2000 (USWRC 1979). This will be a consequence of increased demand for agricultural products, which is expected to result in a 30% increase in crop and livestock production. Factors leading to increased demand for agricultural products include an increase in U.S. population, a greater per capita use of crops grown for food, and more food exports (Pimentel *et al.* 1982).

Far more serious demands may be placed on U.S. water supplies. The production of liquid and gas synfuels from shale oil and coal will require large amounts of water. Pimentel *et al.* (1982) estimate water demand could rise 8–64%, depending on how much of our fuel needs were met by synfuels. This demand is not expected until 2000 and after. Much of this increased water need is expected in an area already beset with water deficits, the West.

Shortages of water will impact heavily upon the use of irrigation. Currently the production of irrigated crops is already costly because of the energy needed for lifting and distributing water (see Chapter 3). For example, irrigated corn in Nebraska easily requires three times as much production energy as rain-fed corn (Pimentel *et al.* 1982). In Arizona energy costs per acre for crop production are five times the national average, with 97% of the costs related to irrigation. Other examples were cited in Chapter 3. Energy costs for pumping irrigation water rose from $350 million to $1 billion during the period from 1973 to 1977 (Sloggett 1979).

As costs rise, the profit on irrigated crops will drop. Competition from rain-fed crops and depletion of groundwater will force the abandonment of some irrigated crops, and a return to dryland farming. Such trends were observed as early as 1977 in parts of Texas and Arizona, where some crops (e.g., alfalfa, barley, sorghum, and wheat) produced losses (Larson and Fangmeier 1977; Patton and Lacewell 1977; OTA 1982). Such trends continue and will likely accelerate in the future. For example, Weigner (1979) pointed out that 41,000 ha (101,270 acres) of irrigated farmland was dropped from production in Pinal County, Arizona. Losses of 1214 ha (3000 acres) have occurred in New Mexico, and another loss of 2388 ha (5900 acres) is expected by 2000 (OTA 1982). Much of the present loss has been with low-value crops, such as alfalfa and other forages.

Larson (1981) predicts, based upon overwithdrawal of the Ogallala aquifer, that reversion to dryland farming will occur on some 1.4 million ha (3.5 million acres) of irrigated cropland in the Great Plains. Similar elimination of irrigated farmland in the Texas panhandle is predicted by Young and Coomer (1980), accompanied by a 60% decrease in crop yields. Another study predicts a loss of 2.1 million ha (5.1 million acres) of irrigated cropland in the Great Plains by 2020 (Pye *et al.* 1983). Again, a reversion to dryland farming is cited as the most immediate solution.

Additional competition for water by synfuel production would aggravate agriculture's water problems. The return on a dollar's worth of water used in the production of synfuels produces a product worth 70 times more than the product resulting from the same water investment in agriculture (Pimentel *et al.* 1982). The future of irrigated agriculture in areas destined for synfuel production appears in doubt. Mitigating circumstances are the uncertainty of synfuel production in view of uncertain oil supplies and the possibility of government intervention to assure provision of water for agriculture.

Additional Problems

Besides groundwater depletion and escalating costs for irrigation, other problems are associated with the overuse and contamination of water by agriculture and other users. These include subsidence, salt problems, and the destruction and stress of aquatic ecosystems. Interestingly, the problem of too much water is encountered with some cropland, thus necessitating drainage. All of these aspects will be examined.

Subsidence

Removal of substantial amounts of groundwater can result in settling of the ground. This phenomenon is termed subsidence and is irreversible. The resulting drop can be far more serious than just damage to buildings, since the resulting stress may lead to faulting, such as occurred in the San Joaquin Valley (Kreitler 1977).

Subsidence in the San Joaquin Valley where groundwater has been withdrawn some 40 years for irrigation, has been particularly severe. Land has dropped in 6475 km^2 (2500 mi^2). In this time drops as much as 6 m (20 ft) and annual rates as high as 0.3 m (1 ft) have been observed, with substantial damage done to agricultural facilities such as buildings, irrigation systems, water-well casings, drainage structures, and flood control structures. Flow directions have been reversed in irrigation canals, and major repairs needed to continue the production of irrigated crops. Costs attributed to repairing subsidence damage in the Santa Clara Valley alone were estimated at $15–$20 million (OTA 1982).

Other areas experiencing subsidence include the San Jacinta Valley in California and the Texan coastal area near Houston and Galveston. Measurements in the San Jacinta Valley indicate that some 13,986 km^2 (5400 mi^2) have dropped 0.36 m (1.2 ft) annually for the last 48 years (OTA 1982).

Correction of the problem is not possible, since subsidence is irreversible. At best the process can be stopped or slowed. Return of water to the aquifer through pumping can stop subsidence. However, this is not practical in view of water scarcity in areas experiencing subsidence. Slower or more widely dispersed removal of groundwater will decrease the rate of subsidence. Again, this may not be easy with limited alternative water resources. The only viable alternatives may be the conversion to dryland farming, the abandonment of all agricultural acticities, or the use of more efficient irrigation. More efficient irrigation will be discussed later in this chapter.

Saltwater Contamination

Heavy withdrawal of groundwater not only leads to subsidence, but also in some instances results in saltwater contamination of the fresh water in the aquifer. Fresh water, being less dense, overlays salt water when present. Such situations are common with aquifers near coastal parts of the United States but also can be found in some inland aquifers.

Heavy pumping of fresh water allows migration of the salt water toward the wells. If the area already has a salinization problem because of poor soil drainage, this condition is further aggravated. Even if salinization is not present, well water, because of encroaching salt water, will become unsuitable for domestic use and ultimately agricultural use.

Saltwater contamination has been observed in some areas where groundwater is being used for irrigation (USWRC 1979). These areas include Dimmit and Zwala counties in Texas, which are served by the Carrizo aquifer, and the part of an aquifer north and east of Roswell, New Mexico. Other areas of saltwater intrusion are coastal areas of California, Florida, Georgia, Louisiana, and Long Island, New York (Anderson and Berkebile 1976; Hammer and Elser 1980; West 1980).

In some instances saltwater contamination is not due to heavy pumping but to improper construction and negligence of wells. As such correction is possible without reducing water removal. Specific examples that lead to saltwater intrusion include active wells that are improperly constructed and incorrectly maintained, or inactive wells that were not sealed and plugged properly. Some of the saltwater intrusion into the Carrizo aquifer may result from old well bores (OTA 1982).

Another source of saltwater infiltration is saline surface waters. Recycling of irrigation water often results in increased salinity. Aquifers then become contaminated by infiltration of saline surface water, especially when over-withdrawal has caused the water table to drop below streams receiving agricultural drainage. Areas of concern at this time are the Trans-Pecos region of Texas and the San Joaquin Valley (USWRC 1979). Both of these areas have overwithdrawn groundwater underlying rivers receiving agricultural drainage and recycled irrigation water.

Preventing saltwater contamination is not easy. Removal of groundwater needs to be slowed or even stopped. Since alternative water resources are not readily available, this may require the cessation of agricultural activities or a return to dryland farming. If irrigation continues, but at a slower pace, a more efficient method (e.g., drip irrigation) may allow maintenance of productivity. Drip irrigation will be covered later in this chapter. Use of recycled irrigation water may increase, but disposal or reclamation systems must be utilized to prevent surface water and ultimately groundwater contamination.

Reduced Surface Water Flow

The list of problems related to heavy pumping of groundwater goes beyond subsidence and saltwater contamination to include reduced flow of surface waters. Normal discharge of groundwater occurs through springs and seepage along the sides of and under streams. Excessive withdrawal of groundwater can lower the level of the water table. A consequence would be reduced or eliminated discharges at springs and declines in stream flow. The latter occurs when a stream no longer receives groundwater seepage, but still infiltrates water to the aquifer below. If this happens, the surface water available for agriculture and other consumers is reduced (OTA 1982).

This makes for a rather vicious circle. Reduced surface water availability forces areas with marginal water supplies to pump groundwater. This increases water mining and raises energy costs, since pumping requires energy, whereas surface water diversion usually operates on gravity flow.

Solutions are not easy. They are essentially those discussed already for saltwater contamination.

Groundwater Contamination

One form of groundwater contamination resulting from agricultural activity, saltwater contamination, has been discussed. Degradation of groundwater also occurs from contamination by animal wastes, fertilizers, and pesticides.

These pollutants are carried in runoff water and drainage water, until they eventually enter the groundwater. The overwithdrawal of groundwater lowers the water table, thus enhancing the infiltration of poor-quality surface water. Groundwater pollutants are troublesome for aquatic ecosystems and public health. Some, such as salts, can make the water unsuitable for agricultural use. Economic factors are linked to the environmental hazards. Costs for treatment of potable water may double and even triple when pollutants must be removed (Clark 1979; EPA 1980).

Pesticides

USDA estimates place the number of marketed pesticides at 1800, of which 1.25 million tons will be used in U.S. agriculture by 1985. Only about 1–2% hit the intended pest and as much as 5% of some pesticides enter the U.S. water supply (OTA 1982; Wauchope 1978).

Some pesticides have been studied enough to document that they can appear, and even persist, in groundwater. Contamination of groundwater by pesticides has been reported in Arizona, California, Maine, Massachusetts, New Jersey, New York, and Texas (Pye et al. 1983; OTA 1982). For example, DDT has appeared in groundwater in Texas, arsenate in Maine, and chlorinated hydrocarbons in Massachusetts (OTA 1982). Toxaphene (a chlorinated hydrocarbon insecticide) and fluometron (a substituted urea herbicide) were monitored in a field study by LaFleur (1973). These compounds took 2 months to infiltrate into the groundwater and persisted during the 1-year monitoring period.

Knowledge about contamination of surface waters by pesticides is somewhat better, but still incomplete. The variables involved in contamination of surface waters through pesticide runoff include the type and amount of pesticide, the area of application, and rainfall timing. Accurate estimate[] actual field inputs into waterways are possible, but knowledge about the impacts of these inputs leaves much to be desired. This results from the scarcity of data concerning dilution dynamics, the sorption and exchange of pesticides among the various soil particles, and pesticide effects upon aquatic ecosystems (OTA 1982; Wauchope 1978). Some insecticides in surface water can lead to fish kills or may destroy part of the fishes' food chain (NAS 1974). Toxic residues from DDT, toxaphene, dieldrin, and aldrin have been reported in freshwater fish and waterfowl (CEQ 1982).

Degradation of pesticides occurs more readily in surface waters than in groundwater. Some organic chemicals that are readily degraded are removed near or before the water enters the aquifer. Some organics may be adsorbed or absorbed by mineral materials in the aquifer. This may cause the accumulation of some organics, while others may travel through the aquifer at rates slower than those not affected by sorption. Those that move slowly or not at all may be more susceptible to microbial degradation, assuming the pH and temperature are suitable.

While surface waters are known to contain complex microbial ecosystems, it was not until recently that groundwater was found to contain such ecosystems. These groundwater microbial ecosystems are dark and oxygen poor, thus anaerobic organisms dominate. Currently, little is known about degradation of organics in aquifers and methods of assessing it, but some pesticide degradation seems probable (McCarty *et al.* 1981, Tangley 1984). Considerable research remains to be done.

Several measures can lead to a reduction of pesticides in surface and groundwaters. The most direct way is eliminate pesticides, as organic farmers do, and instead use biological and cultural methods to control pests and weeds. Short of totally eliminating pesticides, a compromise is to minimize pesticide usage, carefully control applications and timing, use the least-persistent materials, and combine chemical with natural controls and selected managerial techniques—in other words practice integrated pest management. Both of these approaches are discussed in Chapter 9.

Contamination from certain pesticides is reduced or stopped by the adoption of techniques that prevent soil erosion, such as conservation tillage and no-till. This approach works for pesticides that have limited solubility or are tightly sorbed by soil particles. Such pesticides include endrin, paraquat, toxaphene, and trifluralin (OTA 1982). However, because the bulk of pesticides are carried in the water portion of runoff, the volume of which far exceeds the volume of sediment, practices that reduce runoff volume are most important for reducing pesticide pollution. Such practices would include terraces, contour tillage, and cropping systems that maintain high levels of crop residues and keep the soil continuously covered. These practices, discussed in Chapter 7, include rotations and cover crops.

ge and no-till do result in reduced runoff volume, but
er amounts of pesticides. As discussed in Chapter 7, the
pollution appears increased with this form of tillage. The
r the erosion reduction benefits outweigh the pesticide
ly unresolved (OTA 1982), but appears to lean toward

Fertilizers — use leads to nitrate pollution of water

The effects of nutrients upon aquatic ecosystems are known better than
are the effects of pesticides. Nutrients, especially nitrogen and phosphorus,
lead to entrophication of bodies of water. These two nutrients accelerate algal
growth; in turn, the death of the increased algal mass leads to oxygen depletion as oxygen-consuming water microorganisms consume the algae.
Eventually, fish die and the recreational value of the water is diminished.
Runoff from agricultural areas can contain two forms of nutrients: those
dissolved in water and those bound to soil particulate matter. Nitrates are
found mostly in the water portion, and organic forms of nitrogen and inorganic forms of phosphorus are associated with the sediment.

Nitrates in runoff can ultimately appear in groundwater. Runoff of nitrate
from chemical fertilizers is enhanced by certain conditions: high rates of
fertilizer usage, sandy soils, shallow-rooted crops, and heavy rainfall or irrigation (Singh and Sekhon 1978). Other sources of nitrate in runoff are improperly applied manures, sewage sludges, and other organic fertilizers. This
nitrate loss occurs when organic fertilizers are not incorporated into the soil
and runoff appears. Manures or other organics applied to frozen or snow-
covered fields can also lose 10–20% of their nitrogen and phosphorus from
rain or melting snow runoff (Stewart et al. 1976).

Contamination of groundwater with nitrate can be a serious health problem, since nitrate sensitivity (methemoglobinemia) occurs in infants up to 3
months of age. The actual extent of fertilizer nitrate in groundwater is not
resolved (OTA 1982). Still Pye et al. (1983) report that the fertilization of
agricultural land ranks as the second most important source of groundwater
contamination in the Southeast and Southwest. In the Northwest and Northeast, the rank is third and fourth, respectively. California reports nitrates from
agricultural activities as the second most frequently cited source of groundwater contamination. In Texas a jump of over 300% in nitrate levels of the
Seymour water-bearing formation is blamed upon fertilizers. One last concern about nitrate is that continued drinking of nitrate-contaminated water
may lead to the formation of carcinogenic nitrosamines.

Whatever the nitrate source, a number of approaches are possible for
minimizing nitrate pollution of water. One approach is to minimize runoff

(see Chapter 7), which is desirable anyway as a component of erosion control. While the actual difference in nutrient runoff between conservation and conventional tillage is unclear, conservation tillage results in a reduction of nutrient pollution associated with sediment; solubilized nutrient pollution may or may not be less.

However, the use of sod and cover crops, as done by organic farmers, has reduced nutrients in runoff. Losses of nitrogen and phosphorus in runoff were reduced three to six times over those losses observed with continuous corn, when a corn–wheat–clover rotation was utilized. Likewise, a ryegrass cover crop planted to corn reduced nutrient losses in runoff considerably (Stewart *et al.* 1976).

Other practices are also effective in reducing nitrate pollution that occurs through leaching of nutrients. These, and the previous practices, while not exclusively the practices of organic farmers, are heavily used in organic agriculture. Excessive fertilization is avoided. Nitrogen is supplied mostly in the organic form, thus nitrate is produced and released slowly by microbial action, and growing plants use it quickly, leaving little or none for leaching losses. The organic approach is accomplished by using organic fertilizers, such as manures and sludges, green manures, and plowing under of winter cover crops.

The contribution of soil fertility to nitrate pollution is also minimized with practices utilized by organic farmers for cutting fertilizer costs. The rotation of leguminous crops that need little fertilizer with those that need high levels of nitrogen, such as corn or wheat, reduces the long-term average levels of nitrogen available for leaching. Alfalfa appears to be particularly valuable for reducing leaching because of its deep-rooted habit. Winter cover crops help further to reduce nitrate leaching by utilizing the nitrate pool in the nonproductive part of the growing season, and their uptake of water reduces water availability for leaching in the winter. Forage crops, such as oats, timothy, and rye, reduce leaching losses up to one-half. Thus, a pasture–crop rotation would reduce losses through leaching (USDA 1980).

Another factor that is probably important in reducing the contribution of soil fertility to nutrient pollution is the nutrient balance. Low amounts of phosphorus and potassium, relative to nitrogen, lead to accumulation of unused nitrate in the root zone. After harvest this nitrate pool can be easily leached. While excessive nitrogen fertilization can be a problem, so too would be underfertilization with phosphorus and potassium (Singh and Sekhon 1978).

Animal and Human Wastes

The agricultural use of manures and sewage sludges can pollute groundwater, primarily with nitrate. They can also be a source of pathogens, which can enter the groundwater under some conditions (Gerba *et al.* 1981). Management procedures to minimize groundwater contamination by pathogens were discussed in Chapter 7.

Much of the problem results from confined animal feedlots and their disposal facilities, septic tanks, cess pools, and poorly managed sewage disposal facilities. The extent of the problem is not fully known, but in a survey of some ten states, eight indicated that human and animal wastes were among the top three contaminants of their groundwater. These states included Arizona, Connecticut, Idaho, Illinois, Nebraska, New Jersey, New Mexico, and South Carolina.

Proper management of manure can minimize pollution problems. Since this was covered in Chapter 7, only a brief mention is necessary. Animal feedlot operators must carefully control and manage the stock-rate and density of animals. Siting restrictions and proper management practices must be observed. These practices pertain primarily to water containment and diversion, facility operation, and waste storage and disposal. Manure storage and applications on the farm must also be managed correctly if pollution is to be minimized. Covered storage, timely applications, and appropriate loading rates are essential.

Other Wastes

Nonagricultural wastes also contribute significantly to contamination of groundwater. Sources of these include hazardous waste disposal sites (landfills and surface impoundments), injection wells, oil and gas wells, and underground storage tanks. At least 33 toxic organic compounds have been identified in groundwater. Although the primary concern is the health hazard posed for drinking water, the quality of agricultural water is also decreased. The extent of the problem and long-term effects on agriculture are not known.

Since this book is concerned with agricultural groundwater contaminants and known agricultural problems, it is not within our scope to examine these other serious hazardous wastes and their pollution of groundwater. Some of these organic toxins do arise from chemical manufacture of pesticides. Interested readers are referred to much more extensive coverage of these nonagricultural water pollutants by Epstein *et al.* (1982), Pye *et al.* (1983), and Tangley (1984).

Salinization

The presence of excessive salts in soil need not arise from saltwater contamination as discussed previously. This condition may also arise in soils having a drainage problem, which is aggravated by improper irrigation. Salt accumulation under these conditions is referred to as salinization.

As irrigation water is evapotranspired, salts are left behind in the surface soil. These salts increase in concentration and, if uncorrected, reach a concentration that reduces crop productivity. Poor drainage and irrigation water contaminated with salt water or having high levels of soluble salts from fertilizer runoff accelerate salt accumulation.

At this time no official account of how much U.S. irrigated cropland is undergoing salinization is available. Estimates place the figure at 25–35% of the irrigated cropland in the West. The most severe area is probably the western side of the San Joaquin Valley in California (OTA 1982). In this area saline irrigation water is accumulating beneath the surface. This is producing a salty, subsurface water table. At this point the saline water has invaded the root zone of some 161,943 ha (400,000 acres) and is leading to annual crop losses valued at $31.2 million (Sheridan 1981). Unless corrected, the salinized area is expected to be 283,401 ha (700,000 acres) and 404,858–809,717 ha (1–2 million acres) by 2000 and 2080, respectively. The value of crop losses will jump to $321 million by 2000.

Problems such as this are not easily nor cheaply corrected. Improper drainage corrections could cause saline contamination of groundwater and surface waters, and subsequent damage to surface-water ecosystems. Poorly directed drainage could raise saline levels of irrigation water elsewhere. Drainage costs, if the water were piped to a safer disposal area (Pacific Ocean) would run about $185/ha ($75/acre) annually (Sheridan 1981).

An alternative to transporting the saline drainage water elsewhere would be to use it on-site for irrigation of salt-tolerant crops. Such crops are discussed later in this chapter. This would reduce the amount of drainage water, leaving only the most saline portion for disposal at external sites. Problems arise here in getting farmers to participate in changes in farming practices (OTA 1982).

Another salinization problem has been observed in parts of Montana, North Dakota, Oklahoma, South Dakota, Texas, and Wyoming. Farmers in these areas often alternate strips of wheat and fallow. The use of the summer fallow is to conserve moisture in the land strips for later crop use. Salts are solubilized when the saved water becomes excessive. This region is sometimes underlaid by shale, which results in the formation of "perched" saline water tables. The saline water seeps out at the bottom of slopes, creating saline seeps (Fig. 8.1). These are unproductive, swamplike areas, some of which approach 81 ha (200 acres) in size. Some 810,000 ha (2 million acres) are affected by saline seeps in the Northern Plains of the United States (OTA 1982; Brown et al. 1983).

The problem of saline seeps can be alleviated with a crop management system termed "flexible cropping." In this system water conditions are monitored closely, and the data on stored soil moisture serves as the determining factor in choosing the crop to be planted. Such crops include alfalfa, safflower, and sunflower. Conventional crops, such as wheat or barley, may

FIG. 8.1. Saline seeps are especially troublesome on summer fallow land in the Highwood Bench area of Montana.

Courtesy U.S. Congress Office of Technology Assessment

be alternated. Cropping is rotational, but continuous until the water levels are reduced. If reduction is too low, the land is fallowed to conserve water. Such an approach has worked well for saline seep problems in Montana's Highwood Bench, an area near Fort Benton (OTA 1982). More details and computer programs to estimate crop yields with flexible cropping are available (Brown *et al.* 1983).

Drainage

Certain areas suffer from water shortages, not water excesses. Soils in these areas are so wet that it limits their productivity. Estimates place the total area of wet soils in the United States at 109 million ha (270 million acres), of which 40% have wetness that seriously affects crop production (OTA 1982).

The most wet soils are found in the Southeast, which has 17% of the total wet soils. The Corn Belt, Lake states, and Delta states have 16.5, 16.4, and 14.6%, respectively. The other U.S. regions have considerably less: Southern Plains (8.9%), Northeast (7.7%), Appalachian (6.8%), Northern Plains (5.4%), Mountain (3.6%), and Pacific (3.1%).

These wet soils require drainage for crop production. The beneficial effects of drainage can be summarized as follows. In wet soils the micro- and macropores are filled with water, thus the soil is poorly aerated. This condition is detrimental to root development because oxygen is required for root respiration. Drainage removes some water, so the soil becomes aerated and root development improves. The less wet soil also can be worked more easily with farm machinery and warms up faster in the spring, making it possible to start crops 1–2 weeks earlier.

Other benefits of drier soils include reduced water erosion, reduced health hazards, and more efficient disposal of organic wastes. Because drier soil has improved water infiltration, surface runoff and erosion is less. Health hazards from diseases associated with mosquitoes and flies are reduced, since drier soils are not suitable for breeding. Drier soils also provide the oxygen and temperature needed for microbial decomposition of organic matter. An added plus is conservation of nitrogen; drier soils do not provide the cool, wet environment needed for denitrification bacteria.

The methods of soil drainage are summarized well by Troeh et al. (1980). The main area of concern centers on the lack of research and data (OTA 1982). This is a drawback when one must produce efficient design procedures, determine drainage system lifetime, conduct system maintenance, and combine drainage with new cropping technologies, all of which must allow for maximal crop productivity. A second problem is that older farm drainage systems may need repair now in order to prevent eventual failure. However, adequate information may not be available to the unsuspecting farmer.

Conservation of Water

A number of practices are available to the farmer who wishes to conserve water. These management methods can be used to lessen water loss from runoff, evaporation, deep percolation, irrigation, and stored soil water. Practices are also available to maximize the efficiency of irrigation and use of stored soil water, and to make use of water not suited to agricultural crops.

Control of Runoff Losses

A number of practices that control soil erosion caused by runoff have a secondary benefit: They increase the amount of water available for current crop use and storage in the soil. These practices include contour tillage, terraces, crop residues, and water spreading. These practices have been covered in Chapter 7, except for water spreading, and are detailed by Troeh et al. (1980), so coverage here will be brief.

Contour Tillage

This management practice improves the trapping of water, thus allowing longer time for infiltration. Research shows that contour tillage can reduce runoff and increase yields from the extra water (Fisher and Burnett 1953; Ripley et al. 1961). However, not all farmers may benefit. Increased usage of runoff may cause decreased water for farmers who draw upon surface water bodies where runoff provides part of the water input. This problem has been recognized in certain irrigated areas east of the Rocky Mountains (Troeh et al. 1980). Research is needed to establish whether the benefits of reduced soil erosion and increased water conservation on slopes outweighs the loss of some irrigation water.

Terraces

Not all terraces increase water storage. Those terraces that do are designed to hold water until it can soak into the storage. The most effective types are level ridge-type and conservation-bench terraces. These also have the added advantage of controlling water erosion. Design criteria and site restrictions can be found in Troeh et al. (1980).

Crop Residues

The most commonly used practice to maintain organic matter in soil and to control runoff, crop residues also offer the advantage of water conservation. Standing crop residues conserve water through increased trapping of rain and snow, primarily by increased water infiltration. When soil is frozen, the residue helps to hold snow from blowing away, which melts and infiltrates when the soil thaws. For optimal results the crop residue or cover must be maintained to prevent bare, unprotected soil. This involves cover crops and rotations as discussed in Chapter 7.

Water Spreading

The technique of water spreading involves diversion of surface runoff to selected sites where the water infiltrates and is stored in the soil for crop production. This technique is used in Israel in arid portions, but is presently not practiced much in the United States. Details can be found in Medina (1976).

Control of Evaporation Losses

Losses of soil water through evaporation are especially severe in drier areas of crop production. As much as three-quarters of the annual precipitation can be lost through evaporation. Various practices have been used to reduce evaporative losses. The most successful practice is the use of mulches, which also can help control weeds and increase soil temperature. The latter can increase early growth.

The only mulch used on large agricultural operations is black polyethylene plastic. It has been used mostly to control weeds and permit earlier harvests; its application can be mechanized. A bibiolgraphy on synthetic mulches is available (Hopen and Oebker 1976).

Reduction of Losses from Deep Percolation

Soils in arid areas classified as sands are unable to hold large amounts of water in the root zone, and losses from percolation are substantial. Horizontal asphalt barriers placed at depths of about 50–60 cm (19.7–23.6 in.) have increased water storage in root zones of sandy soils (Troeh *et al.* 1980). Chemical amendments, such as "super-slurper" (water-holding starch copolymer) and zeolite minerals, also show great promise for increasing water retention (OTA 1982). These practices have seen little to no use in U.S. agriculture.

Conservation Irrigation

As available irrigation water declines in quantity and quality, the concept of conservation irrigation will become prominent. There will be a move away from inefficient forms of irrigation to more efficient types. Under careful management, efficiency of water use reaches 60% for surface irrigation, 75% for sprinkler irrigation, and 90% for trickle or drip irrigation (Troeh *et al.* 1980). The latter form is most efficient since it supplies water frequently, but slowly, to the soil near the plants, which infiltrates directly to the root zone. The amount applied is approximately just sufficient to the plant's needs. A good description of drip irrigation (Fig. 8.2) is available from the University of California (1979). The latest on energy and water conservation with sprinkler irrigation can be found in Pair (1983). A recent review on drip irrigation also is available (Elfving 1982).

A number of advantages can be cited for drip irrigation (OTA 1982). Water conservation with reasonably managed drip irrigation is at least 15–30% better than that with furrow or sprinkler irrigation. Conservation of fuel and fertilizer is also noted; weeds are significantly reduced, thus labor and herbicide savings are possible. Seedling mortality is reduced and greater uniformity of crops is observed; often increased yields are found. Erosion control is such that drip irrigation can be used on steep terrain where other irrigation is not possible.

FIG. 8.2. Drip irrigation is used mainly on vegetables, fruits, and container stock in the United States. Research indicates it is feasible for row crops, such as corn and cotton, but improvements are needed if drip irrigation is to be economically competitive for row crop use.
Courtesy American Society for Horticultural Science

The leading state in drip irrigation is California, where it is practiced on some 123,482 ha (305,000 acres). The remaining 76,518 ha (189,000 acres) of drip irrigation are scattered in some 30 states (Howell 1981; OTA 1982). Increased use is likely as water supplies dwindle in arid or semi-arid regions, and growers in wetter areas seek to reduce economic risks associated with extensive or seasonal drought.

The main drawback is that initial costs for drip irrigation are more than those for furrow or sprinkler forms. Drip irrigation has been adopted mostly for high-value crops, even though it is feasible for alfalfa, cotton, feed corn, sorghum, and wheat. A partial listing (OTA 1982) indicates that drip irrigation is used with some 22 fruits, 12 vegetables, five nuts, numerous ornamentals, Christmas trees, shrubs and turf. Many home owners are also adopting drip irrigation. This form of irrigation is also frequently seen in greenhouse production.

Drip irrigation systems require good maintenance for continued efficiency. Lines must be flushed and emitters cleaned. Emitters are easily clogged by chemical deposits resulting from soluble fertilizers added to water or natural contaminants (e.g., debris, soil particles, bacteria, algae, and mineral salts). Plant roots may also clog the emitters. Such maintenance is labor intensive, especially when emitter inspection is considered. Deposits from saline water are quite troublesome.

Animals (wireworms, mice, coyotes, rabbits, and others) may damage the plastic lines of drip irrigation systems. This problem can be minimized by placing as much of the system as possible underground, or using plastics that are less attractive to animals.

Despite these drawbacks, the increasing costs for and declining availability of irrigation water will continue to encourage interest in and adoption of drip irrigation. Installation costs are high: $1235–1729/ha ($500–$700/acre). But the benefits cited (Anon. 1983; O'Dell 1983) in studies with peppers and tomatoes include considerable water use reductions (60% less than furrow irrigation), earlier maturation (2 weeks), larger yields (up to 40%), and larger grade sizes.

Recent developments also show considerable promise and will undoubtedly lead to increased use of drip irrigation. High costs slowed the expansion of drip irrigation into moderate-value row crops. However, a traveling trickle irrigation system, which reduces pipe and emitter requirements, is now technologically feasible (Kanninen 1983). The prototype is a lateral move system that pumps water from a concrete ditch running adjacent to the field. Water is distributed from drop tubes as the frame travels across the field. Fertilizers and pesticides can also be metered through the traveling system. It is built with a commercially available frame and one grower has already demonstrated conversion feasibility. Results with cotton crops show increased water application efficiency.

Laser beams and receptors are used to control travel speed and tower alignment. Photovoltaic cells have been used to power the traveling system and could be used to power the pump. This approach offers some energy savings, but has even more value for irrigation of remote sites. Water requirements are determined by a computerized sensing system. The system measures several soil and air variables and can even be tied in with an infrared thermometer. The infrared sensor tests for water stress directly from the crop canopy.

Conventional drip irrigation is also being computerized. Computerized drip irrigation systems were developed and placed into grower use in Israel in the 1970s. They are now appearing in small numbers in Arizona, California, Florida, and Hawaii (OTA 1983; Stout 1984). These systems can increase water and energy use efficiency by 10–30%, minimize labor requirements, reduce water pollution, and lessen drainage problems. The cost payback period, as determined by the Israel experience, is 3–5 years.

Salt-Tolerant Crops

Supplies of irrigation water are becoming saline in some areas, as discussed earlier in this chapter. This problem can be minimized to some extent by growing salt-tolerant plants. This approach would permit water that is becoming saline to remain in crop production longer, and saline drainage water to be recycled rather than disposed.

Three crops have been bred for salt tolerance with some success: barley, tomatoes, and wheat (Epstein *et al.* 1980). Barley is the most salt tolerant. Some developed cultivars can be grown using straight sea water supplemented with nitrogen and phosphorus. The yield with the best cultivar is about half that of barley under normal cultivation. Wheat did not perform as well, but did produce grain with lesser salinity. A cross between a commercial tomato cultivar with a wild salt-tolerant one did produce a cherry-type tomato with some salt tolerance. Much research and breeding remains if salt-tolerant crops are to become commercially viable. The use of tissue culture techniques and genetic engineering may offer hope in this area (OTA 1982a).

Such crops should not be thought of as a panacea. Every effort should be made to prevent water salinity and soil salinization. The present research indicates that yields with salt-tolerant crops cannot be expected to approach those of crops under normal cultivation. Salt-tolerant crops are the avenue of last resort for failed areas of irrigation that have occurred or will occur in the West.

Drip irrigation also shows promise over other forms of irrigation when saline water is used. Other forms of irrigation bring the water into contact with the foliage, leading to high levels of foliar absorption of salts. The net result is a reduction in yield. This problem is minimized with drip irrigation. While more successful, the use of saline water with drip irrigation should be limited to low to moderately saline waters and the more salt-tolerant crops (Elfving 1982).

Over the long run drip irrigation with saline water requires careful attention to salt accumulations in the soil. Leaching with this form of irrigation is minimal. Salinity sensors should be utilized. If rainfall is too low to leach the salts, special irrigation practices must be adopted (Elfving 1982).

We would be wise to look to Israel for improved management practices that minimize salt build-up with drip irrigation. Their growers already use brackish water management practices successfully on vegetables, wheat, and cotton. What is considered highly brackish water in the U.S. is viewed as a low level in Israel (OTA 1983).

Dryland Farming

Another approach to water conservation is simply to return to dryland farming in areas currently using irrigation. Dryland farming uses stored soil water for crop growth and requires a summer fallow period during which water is stored. The stored water is used the following year for crop production.

Summer fallow was widely used in the late 1800s in the Great Plains, but gradually was replaced by irrigation. At present about 15 million ha (37 million acres) of cropland are under summer fallow in the United States. Most of the dryland farming is found in the northern and central Great Plains, where trapping of 30–40% of the precipitation is possible with dryland farming (Troeh et al.1980).

Two problems exist. One, half of the cropland is out of production each year. Therefore, productivity is less than with irrigation. As discussed earlier in this chapter, reversion to dryland farming in the Great Plains would result in a 60% loss in yield. The second problem with summer fallow is saline seeps, which were previously discussed. Lesser problems include the losses of soil by wind and water erosion in fallow areas, and the poor response of corn to dryland farming. In addition, dryland crops usually have lower rates of transpiration than the replaced crops, thus more water might move downward through the soil into the groundwater. Depending on local conditions (e.g., land formation, soluble salts in the soil, and management of fertilizers and pesticides), dryland farming might lead to increased groundwater contamination. More research is needed to assess the risks of these problems.

Still, dryland farming is likely to increase if overwithdrawal of the Ogallala aquifer continues. Crops adapted to dryland farming include wheat, oats, sorghum, and barley. Success and good yields are contingent upon good weed control and maintenance of crop stubble. These practices minimize evapotranspiration. Transpiration by weeds causes the largest losses of stored soil water.

The future in dryland farming may lie in the utilization of food crops suited to the environment. For example, before irrigation the Indians of the American Southwest grew crops. These have been lost or neglected over time as farmers and irrigation transformed the drylands and deserts. A serious effort should be made to bring these crops back and improve them through plant breeding.

One such effort is underway and should be commended: the revival of tepary (Phaseolus acutifolius), which is a legume. This bean was grown by Arizona Indians under natural conditions. Yields are actually decreased with irrigation, so it disappeared. The tepary and efforts to bring it back were explored in a recent symposium (Nabhan 1983).

In addition the U.S. research in other countries where dryland crops are common may provide alternative dryland crops for the United States, and serve as a genetic resource for plant breeding programs (OTA 1983). Leading candidates are beans and cowpeas, which are dietary staples in countries such as Mexico. These legumes offer an economical source of high-quality protein, carbohydrate, and vitamin B. Cooperative research programs exist in this area, but are limited. More funding and increased interest should be directed toward these projects as the conversion to dryland farming increases in the United States.

References

ANDERSON, M. P. and BERKEBILE, C. A. 1976. Evidence of saltwater intrusion in south-eastern Long Island. Ground Water *14,* 315–319.

ANON. 1983. Higher yields with less water. Amer. Veg. Grower *31* (4) 49.

BROWN, P. L., HALVORSON, A. D., SIDDOWAY, F. H., MAYLAND, H. F. and MILLER, M. R. 1983. Saline-Seep Diagnosis, Control and Reclamation. Conserv. Res. Rept. No. 30. U.S. Dept. Agric., Washington, DC.

CEQ. 1980. The Global 2000 Report to the President. Council on Environmental Quality and the Department of State, Washington, DC.

CEQ. 1982. Environmental Quality. 1982. 13th Annu. Rept. Council on Environmental Quality, Washington, DC.

CLARK, R. M. 1979. Water supply regionalization: A critical evaluation. Proc. Am. Soc. Civil Eng. *105,* 279–294.

ELFVING, D. C. 1982. Crop response to trickle irrigation. Hort. Rev. *4,* 1–48.

EPA. 1980. Estimating Water Treatment Costs. Environmental Protection Agency, Washington, DC.

EPSTEIN, E., NORLYN, J. D., RUSH, D. W., KINGSBURY, R. W., KELLY, D. B., CUNNINGHAM, G. A. and WRONA, A. F. 1980. Saline culture of crops: A genetic approach. Science *210,* 399–404.

EPSTEIN, S. S., BROWN, L. O. and POPE, C. 1982. Hazardous Wastes in America. Sierra Club Books, San Francisco.

FISHER, C. E. and BURNETT, E. 1953. Conservation and utilization of soil moisture. Texas Agric. Exp. Stn. Bull. *767.*

GERBA ,C. P., GOYAL, S. M., CECH, I. and BOGDON, G. F. 1981. Quantitative assessment of the adsorptive behavior of viruses to soils. Environ. Sci. Technol. *15,* 940–944.

HAMMER, M. J. and ELSER, G. 1980. Control of ground water salinity, Orange County, California. Ground Water *18,* 536–540.

HOPEN, H. J. and DEBKER, N. F. 1976. Vegetable crop responses to synthetic mulches: An annotated bibliography. Ill. Agric. Exp. Stn. Special Publ. *42.*

HOWELL, T. A., BUCKS, D. A. and CHESNESS, J. L. 1981. Advances in trickle irrigation. *In* Proc. Second National Irrigation Symposium. Amer. Soc. Agric. Eng., St. Joseph, MI.

KANNINEN, E. 1983. Apply water where and when it's needed. Amer. Veg. Grower *31* (5), 17–20.

KREITLER, C. W. 1977. Fault control of subsidence, Houston, Texas. Ground Water *15,* 203–214.

LaFLEUR, K. S. 1973. Movement of toxaphene and fluormeturon through soils to underlying ground water. J. Environ. Qual. *2* (4) 515–518.

LARSON, W. E. 1981. Protecting the soil resource base. J. Soil Water Conserv. *36,* 13–16.

LARSON, D. L. and FANGMEIER, D. D. 1977. Energy requirements for irrigated crop production. *In* Energy Use Management, Vol. I. R. A. Fazzolare and C. B. Smith (Editors). Pergamon Press, New York.

McCARTY, P. L., RITTMAN, B. E. and REINHARD, R. 1981. Trace organics in ground water. Environ. Sci. Technol. *15,* 40–51.

MEDINA, J.   1976.   Harvesting surface runoff and ephemeral streamflow in arid zones. *In* Conservation in Arid and Semi-arid Zones. Food and Agric. Conserv. Guide 3, FAO, Rome.

MILLER, G. T.   1982.   Living in the Environment. 3rd ed. Wadsworth Publishing Co., Belmont, CA.

NABHAN, G. P.   1983.   The desert tepary as a food resource. Desert Plants *5* (1) 1–64.

NAS.   1974.   Productive Agriculture and Quality Environment. National Academy of Sciences, Washington, DC.

O'DELL, C.   1983.   Trickle did the trick. Amer. Veg. Grower *31* (4) 56.

OTA.   1982.   Impacts of Technology on U.S. Cropland and Rangeland Productivity. Office of Technology Assessment, Washington, DC.

OTA.   1982a.   Genetic Technology: A New Frontier. Office of Technology Assessment. Westview Press, Boulder, CO.

OTA.   1983.   Water-Related Technologies for Sustainable Agriculture in Arid/Semiarid Lands. Office of Technology Assessment, Washington, DC.

PAIR, C.   1983.   Sprinkler Irrigation. 5th ed. Irrigation Assoc., Silver Spring, MD.

PATTON, W. P. and LACEWELL, R. D.   1977.   Outlook for energy and implications for irrigated agriculture. Texas Water Resource Institution Tech. Rept. *87*. Texas A & M Univ., College Station.

PIMENTEL, D., FAST, S., CHAO, W. L., STUART, E., DINTZIS, J., EINBENDER, G., SCHLAPPI, W., ANDOW, W. and BRODERICK, K.   1982.   Water resources in food and energy production. Bioscience *32*, 861–867.

PYE, V. I., PATRICK, R. and QUARLES, J.   1983.   Groundwater Contamination in the United States. Univ. Pennsylvania Press, Philadelphia.

RIPLEY, P. O., KABBFLEISCH, W., BOURGET, S. J. and COPPER, D. J.   1961.   Soil Erosion by Water. Canada Dept. Agric. Publ. 1083.

SHEER, D. P.   1980.   Analyzing the risk of drought—The Occoquan Experience. Am. Water Works Assoc. J. *72*, 246–253.

SHERIDAN, D.   1981.   Desertification of the United States. Council on Environmental Quality, Washington, DC.

SINGH, B. and SEKHON, G. S.   1978.   Nitrate pollution of ground water from farm use of nitrogen fertilizers—A review. Agric. and Environ. *4*, 207–225.

SLOGGETT, G.   1979.   Energy and U.S. Agriculture: Irrigation Pumping 1974–1977. USDA Agric. Econ. Rept. No. 436. U.S. Dept. Agric., Washington, DC.

STEWART, B. A., WOOLHISER, D. A., WISTHMEIER, W. H., CARO, J. H. and FRERE, M. H.   1976.   Control of Water Pollution from Cropland. Vol. II. U.S. Dept. Agric. and Environ. Protection Agency, Washington, DC.

STOUT, G. L.   1984.   Now you can computerize drip irrigation. Amer. Veg. Grower *32* (2) 14–15.

TANGLEY, L.   1984.   Groundwater contamination: Local problems become national issue. Bioscience *34*, 142–146, 148.

TROEH, F. R., HOBBS, J. A. and DONAHUE, R. L.   1980.   Soil and Water Conservation for Productivity and Environmental Protection. Prentice-Hall, Englewood Cliffs, NJ.

UNIVERSITY OF CALIFORNIA   1979.   Drip Irrigation. Leaflet No. 2740. Division of Agricultural Sciences.

USDA.   1980.   Report and Recommendations on Organic Farming. U.S. Dept. Agric., Washington, DC.

USDA.   1981.   America's Soil and Water: Condition and Trends. U.S. Dept. Agric., Washington, DC.

USWRC.   1979.   The Nation's Water Resources. Vols. 1–4. U.S. Water Resources Council, Washington, DC.

YOUNG, K. B. and COOMER, J. M.   1980.   Effects of Natural Gas Price Increases on Texas High Plains Irrigation, 1976–2025. Agric. Rept. No. 448. U.S. Dept. Agric., Washington, DC.

WAUCHOPE, R. D. 1978. The pesticide content of surface water draining from agriculture fields—A review. J. Environ. Qual. 7 (4) 459–472.

WEST, D. 1980. Saltwater Intrusion in the Louisiana Coastal Zone. Center for Wetland Resources, Baton Rouge, LA.

WEIGNER, K. K. 1979. Water crisis: It's almost here. Forbes *124* (4) 56–63.

9

Sustaining the
Environment

Agricultural activities have had a definite impact upon the surrounding environment. Effects that have been noted include water, soil, and air pollution. In turn, indirect effects upon animal and even human life have been observed. Unhappily these impacts have resulted in a reduction of environmental quality.

Water quality has declined as a result of agricultural pollutants, such as sediment, nutrients, salts, animal wastes, and pesticides. Air quality has suffered from airborne particulates produced by wind erosion. Pesticides have been harmful to fish and wildlife and have posed a threat to humans when the chemicals become incorporated into the food chain. Animal wastes have contributed to microbial pollution of water, thus posing a disease threat to animal and human life.

An examination of these problems follows, after which agricultural practices that sustain the environment will be covered.

Agriculture's Impact
on the Environment

Water Quality

Pollutants may enter water from a specific defined location (point source) or as a result of runoff from land (nonpoint source). The major cause of nonpoint source pollution is agriculture. Storm runoff, because of present agricultural practices, carries pesticides, particles of soil, nutrients, and organic wastes. Irrigation water return is also a nonpoint source pollutant, when it is saline and contains nutrients.

Sediment forms the largest agricultural pollutant. In fact, the number one pollutant of water by volume is sediment. Erosion of cropland contributes about half of the total sediment, somewhere around 760 million tons a year (USDA 1981). Forestry, construction, and mining contribute the rest.

The detrimental effects of sediment are several fold. It fills in lakes, reservoirs, and rivers, causing displacement of water, which can hinder flood control efforts, increase flood hazards, and even end the usefulness of reservoirs. Eventually, the recreational value and electric power production capacity of bodies of water can be diminished. Costly cleaning and dredging may be necessary.

Sediment pollution also reduces populations of fish and shellfish, and decreases the capacity of waters to assimilate oxygen-demanding wastes. This is caused in part by suspended solids cutting sunlight, thus reducing the growth of aquatic plants and subsequently those animals that feed on plants. Settled solids form an oxygen-demanding sludge. In addition, as discussed in Chapter 8, sediments carry substantial amounts of nitrogen, phosphorus, and low concentrations of certain pesticides. Nutrients lead to eutrophication and sometimes hazardous levels of nitrates. Pesticides can cause fish and animal deaths, and some can be hazardous to humans if they move through the food chain.

The aqueous portion of runoff carries some soluble salts, pesticides, and even microorganisms. Recycled irrigation water, when returned, can raise salinity levels. Soluble nutrients lead to eutrophication; that is, they stimulate growth of algae and water weeds, ultimately causing a reduction of dissolved oxygen. Fish populations and recreational quality decline. Salinity, if high enough, leads to crop damage, unpotable water, and even elimination of some aquatic life. Microorganisms and soluble salts can enter runoff arising from feedlots and sites of manure storage. Some microbes from wastes may be hazardous to humans.

Methods for assessing the deterioration of water quality by these various pollutants are described in an EPA manual (EPA 1982).

Air Quality

Agricultural activities do not lead to serious air pollution. In fact, the two primary air pollutants, oxides of sulfur and nitrogen, arise from industries and utilities that burn coal, oil, or natural gas. These gaseous pollutants can undergo further reactions to produce secondary pollutants. For example, the oxides of nitrogen and sulfur can be oxidized further, and when combined with precipitation, produce the well-known acid rain. Photochemical conversion of nitrogen oxides and unburned hydrocarbons (from internal combustion engines and industry) leads to ozone formation. Acid rain appears to pose no present threat to food crop production, but accumulating evidence suggests a potentially harmful impact upon forest crops (Klein and Klein 1985). Gaseous pollutants, however, are clearly a threat to food production.

Certain types of air pollution pose a considerable threat to agriculture (Kress and Miller 1983; Miller 1983). Sulfur dioxide (SO_2) in low concentrations improves soybean productivity, especially in soils where sulfur availability is limited. Essentially, the sulfur dioxide serves as a nutrient source. However, as SO_2 exposure approaches 10 ppm-hours, yields decrease. Decreases near 20% have occurred as the SO_2 dose nears 20 ppm-hours (Miller 1983).

On the other hand, nitrogen dioxide (NO_2) has little or no effect on soybean yields. However, when combined with SO_2, a powerful synergistic effect occurs. Concentrations of NO_2 that are harmless alone have caused yield losses of soybeans of 20% and even more when mixed with marginally harmful SO_2. Low levels of each (0.13 ppm NO_2 and 0.17 ppm SO_2) when combined reduced seed yields by 18% (Miller 1983).

Ozone also damages plants. Soybeans are highly sensitive (Kress and Miller 1983), and corn, sorghum, and winter wheat are also susceptible, but less sensitive than soybeans. Acid rain appears to pose no threat to these and other crops (Miller 1983).

The main air pollutants arising from agricultural activities are soil particulates and pesticides. Some 30 million tons of particulates are placed into the air annually because of wind erosion of soil. The seriousness of uncontrolled particulate pollution can be vividly demonstrated during the Dust Bowl era of the 1930s. The health of animals and even people were affected by the dust storms. Today, particulates are not nearly as serious, but they do contribute to increased maintenance of machinery, lung and throat irritation, aggravation of respiratory diseases, reduced visibility, and deterioration of buildings. They may even reduce agricultural productivity by decreasing the quality of light reaching plants and by partially obstructing stomates.

Pesticides can also appear in the air through application activities. While airborne, pesticide droplets or particulates can be hazardous if they are inhaled. Once settled from the air, pesticides can be hazardous to fish and animal life, or even to humans, if the chemicals become part of the food chain.

Soil Quality

The most troublesome agricultural pollutants in terms of soil quality are pesticides. Certain toxic wastes, if used as fertilizers or organic amendments, can also be harmful. Both can have an effect upon soil invertebrates and microorganisms.

Responses of soil invertebrates to pesticides are complex, variable, and not thoroughly researched. About the only possible generalization is that pesticide use does change the function and structure of soil communities. Direct and indirect responses by soil invertebrates to pesticides are found. Populations may decrease, but some increase. Some species are destroyed, yet pesticides can accumulate in some organisms without apparent harm. The effects of individual pesticides upon the soil community are understood somewhat, but the effect of multiple pesticides upon the soil community needs much more research.

Pesticides have been known to affect soil microorganisms. This is observed with certain pesticides, but especially with fungicides and fumigants; effects may persist for a long time. The seriousness of microbial changes is evident when microorganisms causing plant diseases or regulating nutrient cycles increase and decrease, respectively.

An area in need of research is whether the use of pesticides damages the soil ecosystem to the extent that crop productivity is lessened over a long period of time. Present data are sparse. Many insecticides and herbicides have the potential to suppress or destroy soil microorganisms if their concentrations become excessive. However, with normal application rates and reasonable mixing of soil through tillage, these chemicals usually fail to reach soil concentrations above 3 ppm. The lack of soil mixing and heavy herbicide usage with no-till and minimal till systems suggests a potential problem needing research. Even if their persistence is limited, the toxicity of degradation products is poorly understood. Ultimately, soil microflora and even crops might be affected, either indirectly through alteration of the soil ecosystem or directly through phytotoxicity.

Pesticides of particular concern include fungicides and soil fumigants. While targeted at pathogens, they usually have a broad spectrum and destroy beneficial microflora. Thus, part of the soil ecosystem is temporarily harmed. Again the long-term effect of residuals is poorly understood.

The present consensus by soil microbiologists is that the short-term effects of pesticides do not warrant their ban. The one short-term problem may be with no-till agriculture, because of extensive herbicide use. Clearly, more research is needed on extensive herbicide use and the long-term effects of pesticides (OTA 1982).

Some forms of organic wastes, when used as fertilizers, have the potential to introduce toxic substances into cropland. These toxins include heavy metals and polychlorinated biphenyls, which are most likely to be present in sewage sludge and organic industrial wastes. This issue is discussed further by Parr *et al.* (1983).

Heavy metals do threaten soil organisms and humans if they become part of the food chain. Soil microbial processes, especially ones of agricultural value like microorganismal decomposition of organic matter and the nitrogen cycle, are inhibited by heavy metals. The level of inhibition is dependent upon the type of heavy metal, its concentration, and the soil type and pH (OTA 1982).

Reducing Agricultural Pollutants

The major share of pollution resulting from agricultural activities can be eliminated. The main pollutant, sediment, arises from erosion. Prevention or control of erosion would accomplish two purposes: sustainability of an agricultural resource and reduction of pollution. The need to control erosion is compelling if agriculture is to be sustained with minimal environmental threat. Practices of erosion control were thoroughly discussed in Chapter 7.

Judicious use and adoption of certain approaches when using organic wastes, fertilizers, irrigation water, and pesticides could minimize the harmful effects of chemicals, nutrients, salts, and pathogens upon the environment. Ways to minimize salinity when using irrigation water have been covered in Chapter 8. Proper handling of manures and other wastes in terms of minimal pollution was discussed in Chapter 7. Since most of the nutrients ending up as water pollutants are carried by sediment (Bradford 1974; Alberts et al. 1978), the most effective practice to minimize nutrient pollution is to control erosion. These practices were discussed in Chapter 7. Ways to control groundwater pollution from nitrates were covered in Chapter 7.

One effective technique not mentioned previously is to use fertilizers judiciously. This has a double benefit: It minimizes pollution and conserves energy. Excess use of fertilizer is not only energy wasteful, but costly and likely to increase pollution. Some additional research is needed to establish ideal quantities of fertilizer based upon crop and soil types. This information must be communicated to the farmer along with the need for frequent soil testing. If farmers realize that money can be saved by judicious use of fertilizers, they are likely to adopt this approach. The secondary impetus would be to promote the fact that the farmer is sustaining agriculture by using less energy and improving the environment.

In the remainder of this chapter, the primary approach to reducing the environmental impact of pesticides is discussed.

Integrated Pest Management

The most reasonable way to minimize pollution by pesticides and yet still retain control over pests is to use integrated pest management (IPM). Interest in and adoption of IPM is such that IPM is becoming a mainstay of sustainable agriculture.

The concept of IPM has been defined by the United Nations Food and Agriculture Organization in the following manner: "A pest management system that, in the context of the associated environment and the population dynamics of the pest species, utilizes all suitable techniques and methods in as compatible a manner as possible and maintains the pest populations at levels below those causing economic injury." Such a management system uses several approaches to pest control. Chemical control with pesticides remains, but as just one part of a coordinated control strategy that also encompasses biological, cultural, competitive, and biorational controls. These components have been well defined by Batra (1982). Two good overviews on IPM are those of Allen (1980) and Frank (1981).

Resource limitation is one of several factors fostering adoption of IPM. Pesticides, being petrochemical products, will become increasingly expensive in terms of both money and energy consumption in the agricultural budget. With its judicious and sparing use of pesticides, IPM appeals to practical-minded farmers and those with concerns for future generations of farmers. Adoption of IPM practices has resulted in real savings, as discussed later in the section on economic benefits.

Environmental concern is another compelling reason for adoption of IPM. Pimentel and Edwards (1982) present a good summary of the well-known problems with pesticides: Some kill beneficial insects or cause other disruptions of natural ecological cycles, and because of their persistence in the environment, some pesticides can kill fish and wildlife and ultimately contaminate the groundwater and organisms in food chains.

One example of a pesticide appearing in the food chain is ethylene dibromide (EDB), which has been shown to be a serious carcinogen in animal tests (Hanson 1984). One use of EDB was for the fumigation of stored grains and milling machinery, and residues of it appeared in some flour, cake mixes, and other grain products. Another use of EDB was as a soil fumigant; as a result it eventually appeared in the groundwater in Georgia and, more seriously, in Florida. Fruit and vegetable contamination also arose, since EDB was used as a postharvest fumigant. Ultimately, the prevalence of EDB contamination caused the Environmental Protection Agency to halt agricultural uses of EDB. Ironically, EDB had replaced a previous soil fumigant, dibromochloropropane, which had been banned a few years before.

Besides environmental and health concerns, another problem is the development of resistance to pesticides. As a result, pesticides become increasingly ineffective, leading to pest control failures. An example is the failure of pesticides to control cotton pests in northeastern Mexico and southern Texas in the late 1960s and early 1970s (Adkisson *et al.* 1982).

A more recent example of pesticide failure is the inactivation of certain pesticides by soil microorganisms (Fox 1983). Microbes in some soils now degrade pesticides so quickly that the efficacy of the pesticide is rapidly diminished. This effect has been noted with several herbicides and some insecticides. Affected crops include corn in the midwestern United States, vegetables, and managed forests. The extent and understanding of this phenomenon is poorly known, since it was only recently discovered.

Other factors that argue for the IPM approach over use of pesticides alone are the decreased crop yields and increased diseases associated with some pesticides (Friend 1983; Hodges and Scofield 1983). For example, yields of lettuce have been decreased by several common pesticides as a result of their adverse effects upon photosynthesis and transpiration. The use of some herbicides has been associated with increased susceptibility to insects and disease. For example, the use of 2,4-D on corn increased the susceptibility to southern corn leaf blight (*Helminthosporium maydis*), corn leaf aphid (*Rhopalosiphum maidis*), and the European corn borer (*Ostrinia nubilalis*).

In spite of widescale pesticide usage, it is estimated that pests cause losses of 33% of annual potential production, worth $20 billion (Batra 1982). These losses lead to an ever-escalating production of new pesticides and increasing costs. Insecticide use has increased 10-fold since the 1940s, yet crop losses from insects have doubled over this same period (Pimentel *et al.* 1983). In contrast the stability and minimal resistance associated with IPM makes it quite attractive.

An economic analysis of corn pest management for the 1980 crop (Hanthorn and Duffy 1983) suggests that pesticides are cost effective, but just barely. For example, farmers who used herbicides realized an estimated return of $1.05 for each $1 spent on herbicides; the return for $1 spent on insecticides was $1.03. Since corn is a major crop, the use of pesticides producing only marginal returns has important implications. However, additional studies are needed to verify this finding. A complication is that 1980 was a drought year. Moisture stress has an effect on yield and the effect of such stress on pesticide effectiveness is difficult to assess. Secondly, pesticide costs were based on national averages, as were returns per bushel of corn. This assumption for pesticide costs may have overstated actual prices paid by farmers. Still, studies like these make IPM look increasingly attractive.

Technological advances with pesticides have helped to make IPM more feasible. While sounding contradictory, two advances have helped to allow better combinations of pesticides with nonchemical controls, such as biological controls. These advances are the development of pesticides that degrade rapidly in the environment and the development of improved, more precisely targeted pesticide application techniques. These improvements increase the chances for biological controls to last long enough to be effective.

A focusing of research interest and money occurred during the late 1970s; by 1982, the majority of the USDA research budget for pest control was directed toward biological and nonchemical controls (Batra 1982). Demonstrations of successful research by the USDA, cooperative extension service, and state agricultural experiment stations helped raise farmers' awareness. Training programs and distribution of information by scientists, extension specialists, and pest management specialists led to acceptance of IPM throughout much of the farming community. Evidence of successful IPM systems in local areas has become the best advertisement for this approach.

Presently IPM is being used to varying degrees on the following crops: apples, citrus, cotton, corn, peanuts, sorghum, soybeans, tobacco, trees, vegetables, wheat, and several minor crops. With some crops, such as corn and wheat, IPM practices are very effective; pest losses increase by only an estimated 1–2%. With other crops, such as apples and potatoes, severe, unacceptable losses occur with IPM methods (Pimentel *et al.* 1983). A good historical approach and overview of IPM is available (Poe 1981).

Practices

It is not within the scope of this book to go into the specific details of various IPM practices, other than in a general sense. Only one specific example, involving an IPM program with cotton, is examined, because of its notable success and recentness. Those interested in a more detailed description of specific practices can consult with appropriate extension specialists and the following sources: USDA (1978, 1980), Flint and Van den Bosch (1981), and Metcalf and Luckmann (1982).

Biological

Pest control in agroecosystems utilizing biological controls is covered by Batra (1982). Based upon her classification, three forms of biological control exist: the classical, augmentative or inundative, and conservative biological controls. Classical biological control utilizes natural enemies of introduced pests. These pathogens, predators, and/or parasites are sought in and introduced from the areas from which the original pest was inadvertently imported. These enemies must be aggressive, host specific, capable of reproduction, and able to achieve sufficient density to become permanent controls of the pest. Classical biological controls have been successful against weeds and insects. Examples can be found in Caltagirone (1981), Frank (1981), and Batra (1982).

Augmentative or inundative biological control involves the periodic release of imported or native enemies after mass propagation (Batra 1982). These enemies, while they may reproduce under some conditions, are not expected to become permanent controls. The pathogens, predators, and parasites do not necessarily have the host specificity that classical biological controls do. Examples cited by Poincelot (1980) and Batra (1982) include several parasitic wasps, an aphid predator (*Chrysopa carnea*), and several insect pathogens. *Bacillus thuringiensis* and *B. popillae* are examples of insect pathogens, in this case against larvae of several moths/butterflies and Japanese beetles, respectively. These particular pathogens are well known commercially.

Conservative biological control utilizes management of all the life forms in an agroecosystem so as to at least conserve, and hopefully to increase, the existing predators and parasites, whether they are native or introduced (Batra 1982). This can be accomplished through use of organic farming techniques, polyculture, and stripcropping. The reasons for failure of these natural controls can often be traced to pesticides that suppressed or eliminated beneficial biota.

Cultural

Practices involved in organic farming appear highly complementary to conservative biological controls (USDA 1980a). Such organic practices as crop rotation, crop spacing, intercropping, mulching, resistant cultivars, and tillage all have some pest control benefits. As observed in earlier chapters, these practices also offer energy savings and soil improvement. Conventional farmers can also manipulate these and traditional practices as part of an IPM program.

Fertilization practices can influence the susceptibility of crops toward diseases and pests (Coleman and Ridgway 1983; Hodges and Scofield 1983). In some instances these problems are associated with excessive fertilization and/or nutrient imbalances. Judicious applications in terms of the amount and proportions of nturients, based upon sound soil-testing programs, can often improve IPM results and save money.

Numerous examples exist, but only a few can be cited here. Many can be found elsewhere (Coleman and Ridgway 1983; Hodges and Scofield 1983). Cotton and rice crops are attacked by increasing numbers and types of pests as fertilization (especially nitrogen) increases; a similar situation occurs with cereal crops and leaf pathogens. Mites on apples increase with nitrogen fertilization, but this increase can be countered by raising phosphorus levels. Aphids on tobacco also rise with increased fertilization. Adequate levels of phosphorus decrease fungal infections of wheat. Squash bug feeding on summer squash seems to be associated with inadequate levels of phosphorus, potassium, and sulfur.

Levels of organic matter in the soil also influence disease and pest susceptibility. Nematode populations and root damage appear to decrease as soil organic matter increases (Van der Lann 1956). Several soilborne diseases (see discussion in Chapter 7) are reduced by incorporation of organic matter into soils (Table 9.1). Organic matter also stimulates microorganisms known to be disease antagonists (Table 9.2). The organic farmer's less than expected problems with insects and disease may well be due in part to his judicious use of organic fertilizers and amendments. Both tend to limit nu-

Table 9.1. *Soilborne Pathogens and Diseases Reduced by Organic Matter*

Pathogen	Crop disease
Aphanomyces cochlioides	Black root of sugar beet
A. euteiches	Pea root rot
Fusarium oxysporum f. *corianderi*	Coriander wilt
F. solani f. *phaseoli*	Bean root rot
Phymatotrichum omnivorum	Cotton root rot
Phytophthora cinnamomi	Avocado root rot
Rhizoctonia solani	Hypocotyl rot of bean
Sclerotinia minor	Lettuce rot, peanut rot
Streptomyces scabies	Potato scab
Thielaviopsis basicola	Bean root rot
	Sesame root rot
	Tobacco root rot

Source: Adapted from Lumsden *et al.* (1983).

trient excesses and imbalances by slow release of nutrients, while maintaining organic matter.

Most often the decomposition of organic matter reduces diseases, but some instances are known where the opposite is true (Lumsden *et al.* 1983). For example, if plant residues previously infected with a pathogen are incorporated into the soil, some diseases may be enhanced. Cotton wilt (*Verticillium dahliae*) and bean root rot (*Fusarium solani* f. *phaseoli*) are examples.

Some cultural practices, such as crop rotations, interplanting, mulching, and time of planting can be complementary to IPM practices (Lumsden *et al.* 1983; Pimentel *et al.* 1983). Rotations of corn with soybean, wheat, and hay are effective for the control of the corn rootworm pest complex. Of these nonhost plants, soybean and wheat are the best choices, as hay crops can introduce several other soil insect pests that attack corn. Rotations of crucifers with grains have reduced several vegetable diseases caused by *Rhizoctonia solani* and *Fusarium* spp., especially in the southeastern United States. In the Northwest, snowmold (*Typhula idahoensis*) can be reduced by rotating wheat with legumes (alfalfa, clover, or peas).

There are some reports that rotations may not be effective for controlling all diseases. Take-all disease of wheat (*Gaeumannomyces graminis*) becomes less severe with extended monoculture, and more severe when wheat is rotated with legumes or hay. As mentioned previously, hay also increases certain corn pests.

Interplanting can reduce some problems, and increase others, as can changing planting times and using mulches (Pimentel *et al.* 1983). Increasing agricultural diversity can reduce insects on certain crops, but the extent to which this occurs is not well documented (Andow 1983). Planting corn early can reduce losses from chinch bug, earworm, fall armyworm, root aphid, rootworm, and southwestern corn borer. However, other pests may become more troublesome and less weed cultivation may be possible. While mulches can control weeds, they may encourage certain pests such as slugs and snails.

Although tillage can provide weed control and the destruction of certain overwintering pests, it tends to increase soil erosion and organic matter oxidation. Minimum and no-till systems reduce the erosion and organic matter problems. The decomposing surface organic matter left in place can control certain diseases, such as stalk rot of sorghum, take-all of wheat, and various soybean rots. However, other disease pathogens are increased: anthracnose, bacterial blight and wilt of corn, and several soybean diseases (Lumsden *et al.* 1983).

Even irrigation can affect diseases (Hodges and Scofield 1983; Lumsden *et al.* 1983). For example, certain diseases are favored by dry soils, and others by wet soils. Well-irrigated tomato plants can have increased rots from anthracnose and *Rhizoctonia*, but less blossom-end rot. Other pathogens that are less prevalent with heavy irrigation include *Fusarium oxysporum* f. *cubense, Phymatotrichum omnivorum, Sclerotinia sclerotiorum,* and *Verticillium albo-atrum.*

Careful control of cultural practices can increase the effectiveness of an IPM system. Much of the organic farmer's success without pesticides may be attributed to the conscious and sometimes serendipitous combinations of practices. Our knowledge in this area is incomplete, but the potential for pest control through proper combinations of cultural and biological controls exists. More research is needed to clarify control of specific pests on various

Table 9.2. *Antagonstic Microorganisms Stimulated by Organic Matter*

Microorganism	Suppressed pathogens or diseases
Actinomycetes	
Streptomycetes	*Rhizoctonia solani*
Bacteria	
Pseudomonas cepacia	*Fusarium solani,*
	Rhizoctonia solani,
	Thielaviopsis basicola,
	Verticillium albo-atrum
Fungi	
Gliocladium roseum	*Gaeumannomyces graminis,*
	Rhizoctonia solani
Trichoderma hamatum	*Rhizoctonia solani*

Source: Adapted from Lumsden *et al.* (1983).

crops through the complementary combination of cultural and biological controls.

IPM with Cotton

Integrated pest management has been quite successful with cotton and has resulted in decreased usage of pesticides, fertilizers, and irrigation. The cotton IPM system, which is discussed by Adkisson *et al.* (1982), is becoming widely accepted not only because of environmental concerns but because it produces greater net returns than other pest control approaches.

The successful use of IPM in cotton is a significant landmark in the development of sustainable agriculture. Nearly half of the insecticides utilized in American agriculture are used on cotton. Substantial reduction of national insecticide use would occur if IPM were adopted extensively by cotton growers. In the area of widest adoption, Texas, insecticide use in the late 1970s was only 12% of its highest use in 1964 (OTA 1979).

Five pests are most common in cotton production: the boll weevil (*Anthonomus grandis*), cotton fleahopper (*Pseudatomoscelis seriatus*), pink bollworm (*Pectinophora gossypiella*), tobacco budworm (*Heliothis virescens*), and bollworm (*Heliothis zea*). The first three are primary pests, those that cause the most trouble and reduce profits. The last two are secondary pests; these types only cause problems when their natural enemies disappear.

As insecticide usage in cotton became common during the 1950s, resistance among the primary pests increased, causing the need for different insecticides and/or increased dosages. These changes decimated the natural enemies of the secondary pests, which soon became essentially primary pests. By the late 1960s and early 1970s, problems of resistance peaked. Most registered insecticides for cotton were so ineffective that cotton production in Texas dropped to one-third of its peak. The stage was set for a new, innovative approach: IPM (Adkisson *et al.* 1982). Ways were sought to minimize insecticide usage and maximize control by natural enemies.

A number of elements were combined to produce a very effective IPM system. Conventional long-season cotton cultivars set bolls during peak times for boll weevils, making bolls very susceptive to damage. Therefore, short-season cultivars were developed; these fruited early enough that the boll's carpel was thick enough to resist attack by the time the boll weevils peaked. Further refinements of these cultivars resulted in resistance to seedling pathogens, immunity to bacterial blight, and moderate resistance to the cotton fleahopper. One particular cultivar also has moderate resistance to the boll weevil, boll worm, and tobacco budworm.

These cultivars, coupled with certain cultural practices, essentially eliminated the need for most insecticides. Shredding and plowing under stalks by mid-September kills most of the diapausing weevils. An insecticide used after harvest further reduces diapausing weevils. Some insectide is used in the spring to control cotton fleahopper and overwintered weevil adults to prevent reproduction. No sprays are needed to control later generations of weevils. This spares the natural enemies of the secondary pests, thus assuring their control. Pink boll worms are controlled by early planting and destruction of crop residues.

The use of a short-season cultivar of cotton also reduces the need for fertilizer and irrigation water. Most cotton farmers in Texas now use some form of this IPM system, which has made it possible for Texas to again produce about 50% of the U.S. cotton crop.

Economic Benefits

Environmental and sociological concerns alone would probably not be enough to cause widespread adoption of IPM in the near future. Fortunately, economic benefits also may occur, thus serving as an impetus for adoption and increased research. The road to complete adoption of IPM as a cornerstone of sustainable agriculture is not fully paved, but certainly has been begun. The actual extent of present usage is not well documented in the literature.

The extent of economic benefits also is not well-documented. However, the few reported cases strongly suggest dollar savings are quite possible. The best documentation is that by Adkisson *et al.* (1982). Use of IPM systems on cotton, as described in the previous section, resulted in a fourfold increase in net returns per acre compared with conventional pesticide systems. This figure was determined with a well-run demonstration on a private farm where conditions were optimized. Average increases in returns on typical commercial farms were not as great, but were certainly notable. Cotton producers on farms in coastal Texas realized increases from $153/ha ($62/acre) with conventional pesticides to $420/ha ($170/acre) with IPM systems. Dryland farmers in Texas had increases of $77/ha ($31/acre). A discussion of the economics of IPM on cotton production in the coastal bend region of Texas is presented by Masud *et al.* (1981).

Another report (Frecon 1983) cites savings of $91/ha ($37/acre) when an IPM program was used on 2470 ha (1000 acres) of apples and 494 ha (200 acres) of peaches in New Jersey. A second report (Anon. 1983) cites the work of Roger Adams at the University of Connecticut and his IPM system for sweet corn.

This sweet corn project ran for 3 years and involved 10 growers with up to 607.3 ha (1500 acres). This is about 40% of Connecticut's corn land. Actual tests of IPM versus standard pest management were held in sections of each grower's field. Traps utilizing blacklight and pheromones were placed for monitoring the main corn pests: corn earworm (also called cotton bollworm, *Heliothis zea*), European corn borer (*Ostrinia nubilalis*), and the fall armyworm (*Spodoptera frugiperda*). Field scouts were also used as monitors. On the IPM plots, pesticide usage declined on the average by 25%; the high was 50%. This decrease in turn had the indirect benefit of reducing related costs for pesticide usage, such as fuel, equipment maintenance, and labor. Less compaction of soil also was observed, since fewer pesticide field passes were needed. In addition, the corn quality was higher on IPM plots as opposed to standard practices.

Extensive reviews containing earlier information on the economics of IPM systems are available (McCarl 1981; Osteen *et al.* 1981). Other analyses of the economic potential and impact of IPM strategies can be found in Araji (1981), Musser *et al.* (1981) and Thompson and White (1981).

Current Research

Several factors will determine the future development and status of IPM: current and future research; sociological, environmental, regulatory, and economic considerations; changes in technology; improvements in education and informational dispersal; and farmer acceptance. A number of problems also exist, which may or may not be resolved. For example, specific IPM systems for crops cannot necessarily be developed on a national scale, since regional differences with pests and other aspects of the agroecosystem exist.

Research Directions

Most research tends to focus on specific crops and the pests and other agroecosystems associated with them. The one exception is the gypsy moth (*Porthetria dispar*), the most serious defoliating insect of hardwood trees in the forests, woodlands, and yards of homes in the eastern United States, especially the Northeast.

The USDA and agricultural research stations in the affected states have waged a massive research effort against this gypsy moth. A summary of all the research targeted toward IPM control of this horticultural and forestry pest is covered in excellent detail by Doane and McManus (1981). A proposal for the use of IPM for the cooperative control of the gypsy moth in the most seriously affected states (Connecticut, Maine, Maryland, Massachusetts, New Hampshire, New Jersey, New York, Pennsylvania, Rhode Island, and Vermont) can be found in USDA (1981a).

Much IPM research is targeted toward specific large production crops. Recent examples include alfalfa hay (Flint 1981), cotton (Adkisson 1981), rice (Khush and Chandhary 1981), soybeans (Kogan 1981; Carnahan 1982), tobacco (Cheng 1981), and wheat (Nalewaja 1981). Crops of horticultural interest are also the subject of recent IPM research (Wearing 1982). These include apple (Brunner 1981), citrus (Elmer *et al.* 1981), *Gypsophola* sp. and chrysanthemums (Price *et al.* 1981), poplar (Filer *et al.* 1981), potato (LeBrun 1984), deciduous fruits (Fisher and Weinzierl 1981; Mathys 1981), turf (Gibeault *et al.* 1981), and woody ornamentals (Miller and Nielsen 1981). Some IPM research has been targeted toward forest crops (Coulson 1981; Schmidt and Wilkinson 1981; Ciesla 1982).

A newly emerging research area is that of urban IPM. The control of insects and rodents in inner cities; insects, diseases and weeds in the turf and gardens of urban and surburban sites; and insects and diseases on city trees and in parks is beset with difficulties. Diversity is a big problem, whether one thinks of the many pest problems or the different people involved. Further complications exist in terms of economics and personal liberties. Recent reviews on urban IPM research include Carnahan (1982a), Levenson and Frankie (1981), and USDA (1981b).

Another concern of IPM researchers is the influence of environmental conditions and climate on the effectiveness of control by IPM. Coverage of this topic can be found in Duniway (1982), Hatfield and Thomason (1982), and Welch (1982).

Newer vectors of control, to supplement or replace traditional insect predators and parasites and fungal diseases, are receiving attention and may well become the focus of new research. Examples include the use of viruses against insects, allelopathic chemicals against insects and weeds, weeds to serve as sites for beneficial insects and the foiling of pests, and pheromones to suppress insects. The use of insect viruses as control agents has been covered in detail by Payne (1982); Miller *et al.* (1983) discuss all microbial controls.

Allelopathic chemicals are natural chemical inhibitors released by plants and microorganisms. These chemicals arise from the leaves, litter, and roots in plants and are released into the soil through leaching, decay, and exudation. They are of particular interest because of their herbicidal properties. Interestingly, some of the crops used in rotation schemes on farms show strong allelopathic traits. These crops include oats, soybeans, sorghum, wheat, rye, barley, and sunflower. These particular crops appear to suppress weed growth for succeeding crops not showing strong allelopathic characteristics. Weed control through the use of annual rotations or companion plantings of allelopathic crops may become a future agricultural practice. An excellent discussion of allelopathic chemicals is presented in Putnam (1983).

Normally, the IPM approach toward weeds has been to reduce their growth or eliminate them. Recently, attention has focused on the use of certain weeds as hosts for beneficial species, as repellants against harmful insects and nematodes, and as decoys or traps for undesirable insects and pathogens. An excellent summary of these approaches can be found in William (1981).

Controlled release of pheromones, natural sexual attracts of insects, to suppress insects has received widespread attention. Many home owners have utilized pheromone traps against gypsy moths and Japanese beetles. Some pheromones have been used successfully with major crops, such as cotton. Detailed, comprehensive coverage of this topic is available (Kydonieus and Beroza 1982).

Future Directions

A number of issues are in need of further resolution. For example, the methodology to predict pest arrivals, population levels, and damage consequences needs refinement. One approach possible is systems modeling, which is based upon mathematical modeling or simulations of crop growth and pest dynamics with the assistance of computers. While theoretical, proper development of models can produce an operational approach fully capable of providing the predictions required for success with IPM. Crop scouting, a practical tool, is used for assessment of crops and pests. It provides an ongoing informational basis for IPM use, as opposed to the previous predictive informational base. Both are needed for good results with IPM systems. Someday, the ongoing information collected by scouts may be transmitted via computer links directly to the farmer. One such program like this almost exists in Kentucky. Almost, because presently the information exists on the computer in the county extension office and the farmer calls in to get it (Brown 1982). The obvious future is computer access directly from the farmer's home computer.

Modeling presently plays some role in IPM. For example, some existing models cover the dynamics of pest populations for specific crops and pests. Covered crops include alfalfa, cotton, forests, peanuts, and soybeans (Ruesink 1976). Some models of plant pathogens (Tummala *et al.* 1976) are available: apple scab (*Venturia inaequalis*), potato blight (*Phytophthora infestans*), and wheat stem rust (*Puccinia graminis tritici*). Hartstack and Witz (1981) discuss modeling of alfalfa weevil (*Hypera postica*), boll weevil (*Anthonomus grandis*), earworms (*Heliothis*), Lygus bugs (*Lygus hesperis*), and mites. To date, models have concerned major pests or crops, where economic losses can be substantial; expansion of models to other crops and pests is needed.

Modeling is also the working tool of system science, which is and will be increasingly applied to agriculture, especially IPM. Agricultural scientists accept the systems concept, which basically seeks to resolve complex agroecosystems into various components and subsystems. Modeling is used to provide a working facsimile of each component. Each component interacts with the other to make up the system. The model describes the component and its relationship to others. These models can be manipulated to predict the behavior of the system, given different inputs. A good discussion of the basic mathematics and computer input of agricultural modeling is available (DeMichele 1975).

A number of factors are involved as components in the system. These include crop development, pests and pest enemies, cultural factors, and several environmental factors (e.g., humidity, nutrients, rain, soil, temperature, and water). These components may either be controlled or uncontrolled factors. Controlled factors can include, for example, cultural practices and cultivar qualities, whereas uncontrolled factors include pests, temperature, and nutrients.

Since uncontrolled factors change with time, they require frequent updating to truly reflect the current status of the system. If done, the model system becomes a working tool for ongoing decision making and adaptibility to changing conditions. Such an approach is called on-line pest management (Tummala 1976).

Used in this manner, system science application to pest management or other agricultural practices offers several uses. It can provide resolution and insights into complex agroecosystems, identify specific areas as factually weak and serve to show research priorities, predict behavior and serve to assess needs of pest management and other cultural practices, and serve to provide data for interdisciplinary research needs.

Much more research is needed on system science for IPM. Presently, large-scale models exist for forestry and cotton (Ruesink 1976), two crops that now face increasingly serious pest threats. Conventional pest management appears less desirable for these crops because of rising costs, increased pesticide resistance, the need for increasingly stronger pesticides at higher dosages, and rising environmental concerns.

The other models discussed earlier are not for the total system, but are for some of the system components. More modeling is needed to complete those systems, and the system science approach needs to be extended to presently uncovered crops. Present crops that depend heavily on modeling in their IPM programs include apple, celery, citrus, corn, cotton, and soybeans (Poe 1981).

Modeling systems also need to accommodate regional and even local differences in IPM programs. This undoubtedly will involve extension agents and some computer technology at the regional and even farm level. Acceptance of computer technology at the farm level will undoubtedly happen at corporate farms and to a degree at single family farms (see Chapter 10). Consultants, who already supply much IPM advice, may also supply computer information at local levels.

In conclusion IPM will remain and expand. More research is needed at the university and state experiment station level, cooperation is needed on the regional level for planning and coordination, and education must supply the professional consultants, knowledgeable extension agents, and aware farmers. The USDA commitment of extensive funds to IPM research, regional programs, and education is encouraging (Kuhr 1981; Batra 1982).

Genetic Resources

Germplasm is a very critical agricultural resource which receives little attention. Crop productivity increases through plant breeding have been and will be dependent upon germplasm resources. The lack of attention results from germplasm's low visibility as a resource. Germplasm is not observable like soil and water, nor is the erosion of genetic resources as obvious a problem as soil erosion or water loss.

We know that successful plant breeding is highly dependent upon the worldwide availability of genetic diversity. Plant breeders use germplasm for a number of plant improvements. One important use is the introduction of cultivars showing first time resistance towards an existing disease or insect. Equally important is the reintroduction of resistance in plants where modified strains of diseases and insects have managed to circumvent existing resistance. Other new cultivars are bred to satisfy changes in cultivation or market demands. Improved production of natural products from crops, such as medicinals or substitute fuels, is also dependent upon the diversity of genetic resources.

Genetic diversity is being lost at an alarming rate outside the United States (OTA 1982a). Losses are especially heavy in tropical forests as a consequence of increased urbanization, industrial development, and agricultural production. Estimates place the current total loss of tropical forest cover at 40% and the annual clearance loss continues at 1–2%. Unless abated, tropical forests will be 75% lost by the year 2000. While assessment is difficult, the remaining tropical forests are extrapolated to contain nearly one half of the earth's plant species. The majority of these species are unknown in terms of identity and value as a genetic resource. Estimates place the ongoing yearly plant extinctions at several hundred in addition to thousands of indigenous wild crop types lost already. Once lost, germplasm is irreplaceable.

Solutions to the problem are not easy, since costs are high and the impacted areas are beyond the United States (OTA 1982a). Clearly germplasm repositories in the United States must be enlarged and broadened. Collection efforts need to be increased. Equally essential is the preservation of natural genetic resources until scientists are able to complete the identification, collection and evaluation of the sources of genetic diversity. Options concerning germplasm have been presented to Congress by the Office of Technology Assessment.

A related threat concerns genetic vulnerability. Our major crops now have a narrow genetic base as a consequence of inbreeding resulting from the needs of farmers, food processors, and consumers. This selection for and national dominance by a few best cultivars for each crop is potentially dangerous. Should a disease mutation arise and infect a major crop, severe damage could result for that crop. The devastating corn blight in 1970 raised the specter of food crop susceptibility to severe disease outbreaks.

Increased genetic diversity in our major crops is desirable and possible, but social, economic and free market pressures complicate the issue. This area clearly needs more attention. At the least (OTA 1982a), an early warning system for the recognition of potential vulnerability of crops should be developed.

References

ADKISSON, P. L., NILES, G. A., WALKER, J. K., BIRD, L. S. and SCOTT, H. B. 1982. Controlling cotton's insect pests: A new system. Science *216*, 19–22.

ALBERTS, E. E., SCHUMAN, G. E. and BURWELL, R. E. 1978. Seasonal runoff losses of nitrogen and phosphorus from Missouri Valley loess watersheds. J. Environ. Qual. 7 (2), 203–208.

ALLEN, G. E. 1980. Integrated pest management. Bioscience *30*, 655–701.

ANDOW, D. 1983. Effect of agricultural diversity on insect populations. *In* Environmentally Sound Agriculture. W. Lockeretz (Editor). Praeger Publishers, New York.

ANON. 1983. Connecticut pest management with IPM. Amer. Veg. Grower *31* (4) 46.

ARAJI, A. A. 1981. The economic impact of investment in integrated pest management. Res. Bull. *115*. Idaho Agric. Exp. Stn.

BATRA, S. W. T. 1982. Biological control in agroecosystems. Science *215*, 134–139.

BRADFORD, R. R. 1974. Nitrogen and Phosphorus Losses from Agronomy Plots in North Alabama. EPA-660/2-74-033. U.S. Environ. Protection Agency, Washington DC.

BROWN, G. C. 1982. Microprocessor-based information management system for an integrated pest management system. Bull. Entomol. Soc. Amer. *28*, 135–137.

BRUNNER, J. 1981. New tactics for apple IPM. pp. 145–150. *In* Proc. Washington State Horticultural Assoc. 1981.

CALTAGIRONE, L. E. 1981. Landmark examples in classical biological control. Annu. Rev. Entomol. *26*, 213–232.

CARNAHAN, W. B. 1982. Soybean IPM. Extension Rev. *53* (2) 6–7.

CARNAHAN, W. B. 1982a. Urban IPM: Blooming across the country. Extension Rev. *53* (2), 12–14.

CHENG, H. H. 1981. Progress in integrated pest management of tobacco insects in Canada. The Lighter *51* (3) 17–21.

CIESLA, W. M. 1982. IPM: New approaches to old problems. Amer. For. *88* (2) 40–44, 51–52.

COLEMAN, E. W. and RIDGWAY, R. L. 1983. Role of stress tolerance in integrated pest management. *In* Sustainable Food Systems. D. Knorr (Editor). AVI Publishing Co., Westport, CT.

COULSON, R. N.   1981.   Evolution of concepts of integrated pest management in forests. J. Ga. Entomol. Soc. *16* (Suppl. 1), 301–316.

DeMICHELE, D. W.   1975.   An evaluation of modelling, systems analysis, and operations reserach in defining agricultural research needs and priorities in pest management. Iowa State J. Res. *49,* 597–621.

DOANE, C. C. and McManus, M. L.   1981.   The Gypsy Moth: Research Toward Integrated Pest Management. Tech. Bull. U.S. Dept. Agric. Forest Service, Washington, DC.

DUNIWAY, J. M.   1982.   Role of biometeorology in integrated pest management: Soil, plant, water relations and disease. *In* Biometeorology in Integrated Pest Management. Academic Press, New York.

ELMER, H. S., BRAWNER, O. L. and EWART, W. H.   1981.   The citricola scale in San Joaquin IPM programs. Citrograph *66* (6) 137–138, 151.

EPA.   1982.   Water Quality Assessment: A Screening Procedure for Toxic and Conventional Pollutants. EPA 600/6-82-004a-c. Center for Environ. Res. Information, U.S. Environ. Protection Agency, Cincinnati, Ohio.

FILER, T. H., JR., SOLOMON, J. D., COOPER, D. T. and HUBBES, M.   1981.   Integrated pest management of poplar species. *In* Proc. IX Intern. Cong. Plant Protection, 1979. Burgess Pub., Minneapolis.

FISHER, G. C. and WEINZIERL, R.   1981.   Integrated Pest Management and Its Potential Application in Small Fruits Production. Annu. Rept. Oregon Horticultural Soc., Vol. 72.

FLINT, M. L.   1981.   Integrated Pest Management for Alfalfa Hay. Statewide IPM Project. IPM Manual Group. University of California, Berkeley.

FLINT, M. L. and VAN DEN BOSCH, R.   1981.   Introduction to Integrated Pest Management. Plenum Press, New York.

FOX, J. L.   1983.   Soil microbes pose problems for pesticides. Science *221,* 1029–1031.

FRANK, J. R. (Editor)   1981.   Integrated pest management: Present and future. Proc. of symposium, sponsored by Weed Control and Pest Management Working Group of the American Society for Horticultural Science. HortScience *16,* 500–516.

FRECON, J.   1983.   IPM saves money. Amer. Fruit Grower *103* (1) 19.

FRIEND, G.   1983.   The potential of an alternative agriculture. *In* Sustainable Food Systems. D. Knorr (Editor). AVI Publishing Co., Westport, CT.

GIBEAULT, V. A., BOWEN, W. R., OHR, H. D., THOMASON, I. J. and CRESS, F.   1981.   Integrated pest management for turf. Calif. Turfgrass Culture *31* (2) 13–15.

HANSON, D. J.   1984.   Agricultural uses of ethylene dibromide halted. Chem. & Eng. News (March 5), 13–16.

HANTHORN, M. and DUFFY, M.   1983.   Returns to Corn Pest Management Practices. Agric. Econ. Rept. No. 501. U.S. Dept. Agric., Washington, DC.

HARTSTACK, A. W. and WITZ, J. W.   1981.   Insect modeling. Agric. Eng. *62* (9) 19–20.

HATFIELD, J. L. and THOMASON, I. J. (Editors).   1982.   Biometeorology in Integrated Pest Management. Academic Press, New York.

HODGES, R. D. and SCOFIELD, A. M.   1983.   Effect of agricultural practices on the health of plants and animals produced: A review. *In* Environmentally Sound Agriculture. W. Lockeretz (Editor). Praeger Publishers, New York.

KHUSH, G. S. and CHANDHARY, R. C.   1981.   Role of Resistant Varieties in Integrated Pest Management of Rice. Extension Bull. No. 162. Food and Fertilizer Technology Center, Taipei City, Taiwan.

KLEIN, D. T. and KLEIN, R. M.   1985.   The growing case against acid rain. Garden *9* (2) 22–27.

KOGAN, M.   1981.   Dynamics of insect adaptations to soybean: Impact of integrated pest management. Environ. Entomology *10* (3) 363–371.

KRESS, L. W. and MILLER, J. E.   1983.   Impact of ozone on soybean yield. J. Environ. Qual. *12* (2) 276–281.

KUHR, R. J.   1981.   Regional planning and coordination of integrated pest management programs. HortScience 16, 514–515.

KYDONIEUS, A. F. and BEROZA, M.   1982.   Insect Suppression with Controlled Release Pheromone Systems. Vols. 1 and 2. CRC Press, Boca Raton, FL.

LeBRUN, R. A.   1984.   Will *B.b.* beat the beetle? Amer. Veg. Grower *32* (1) 8.

LEVENSON, H. and FRANKIE, G. W.   1981.   Pest control in the urban environment. Prog. in Resource Management and Environ. Planning *3*, 251–272.

LUMSDEN, R. D., LEWIS, J. A. and PAPAVIZAS, G. C.   1983.   Effect of organic amendments on soilborne plant diseases and pathogen antagonists. *In* Environmentally Sound Agriculture. W. Lockeretz (Editor). Praeger Publishers, New York.

MASUD, S. M., LACEWELL, R. D., TAYLOR, C. R., BENEDICT, J. H. and LIPPKE, L. A. 1981.   Economic impact of integrated pest management strategies for cotton production in the coastal bend region of Texas. South. J. Agric. Econ. *13* (2) 47–52.

MATHYS, G.   1981.   Implementation of integrated pest management on deciduous fruits in Europe. *In* Proc. IX Intern. Cong. Plant Protection, 1979. Burgess Pub., Minneapolis.

McCARL, B. A.   1981.   Economics of integrated pest management: An interpretive review of the literature. Oregon Agric. Exp. Stn. Special Rept.

METCALF, R. L. and LUCKMANN, W. H.   1982.   Introduction to Insect Pest Manatement. 2nd ed. John Wiley & Sons, New York.

MILLER, J. E.   1983.   Studies reveal effects of air pollution on crops. Logos *1* (3) 20–23.

MILLER, L. K., LINGG, A. J. and BULLA, L. A.   1983.   Bacterial, viral and fungal insecticides. Science *219*, 715–721.

MILLER, R. L. and NIELSEN, D. G.   1981.   Insect and Mite Control on Woody Ornamentals. Ext. Bull. *504*. Ohio State Univ.

MUSSER, W. N., TEW, B. V. and EPPERSON, J. E.   1981.   An economic examination of an integrated pest management production system with a contrast between E-V and stochastic dominance analysis. South. J. Agric. Econ. *13* (1) 119–124.

NALEWAJA, J. D.   1981.   Integrated pest management for weed control in wheat. *In* CRC Handbook of Pest Management in Agriculture, Vol. 3. CRC Press, Boca Raton, FL.

OSTEEN, C. D., BRADLEY, E. B. and MOFFITT, L. J.   1981.   The Economics of Agricultural Pest Control: An Annotated Bibliography 1960–80. U.S. Dept. Agric. Economics and Statistics Service.

OTA.   1979.   Pest Management Strategies: Present and Future Pest Management Strategies in the Control of Sorghum and Cotton Pests in Texas. Office of Technology Assessment, Washington, DC.

OTA.   1982.   Impacts of Technology on U.S. Cropland and Rangeland Productivity. Office of Technology Assessment, Washington, DC.

OTA.   1982a.   Genetic Technology: A New Frontier. Office of Technology Assessment. Westview Press, Boulder, CO.

PARR, J. F., MARSH, P. B. and KLA, J. M.   1983.   Land Treatment of Hazardous Wastes. Noyes Publications, Park Ridge, NJ.

PAYNE, C. C.   1982.   Insect viruses as control agents. Parasitology *84* (4) 35–77.

PIMENTEL, D. and EDWARDS, C. A.   1982.   Pesticides and ecosystems. Bioscience *32*, 595–600.

PIMENTEL, D., GLENISTER, C., FAST, S. and GALLAHAN, D.   1983.   An environmental risk assessment of biological and cultural controls for organic agriculture. *In* Environmentally Sound Agriculture. W. Lockeretz (Editor). Praeger Publishers, New York.

POE, S. L.   1981.   An overview of integrated pest management. *In* HortScience *16*, 501–506.

POINCELOT, R. P. 1980. Plant protection. *In* Horticulture: Principles and Practical Applications. Prentice-Hall, Englewood Cliffs, NJ.

PRICE, J. F., ENGLEHARD, A. W., OVERMAN, Q. W., YINGST, A. J. and IVERSON, M. K. 1981. IPM demonstrations in commercial gypsophola and chrysanthemums. *In* Proc. of the 1980 Meeting, Florida State Horticultural Society.

PUTNAM, A. R. 1983. Allelopathic chemicals: Nature's herbicides in action. Chem. Eng. News (April 4), 34–45.

RUESINK, W. G. 1976. Status of the systems approach to pest management. Annu. Rev. Entomol. *21*, 27–44.

SCHMIDT, R. A. and WILKINSON, R. C. 1981. Prospects for integrated pest management in slash pine ecosystems in Florida. *In* Proc. IX Intern. Cong. Plant Protection, 1979. Burgess Publ., Minneapolis.

THOMPSON, P. and WHITE, G. B. 1981. An Economic Evaluation of the Potential for Tree Fruit Integrated Pest Management in the Northeast. Dept. of Agricultural Economics, Cornell Univ. Agric. Exp. Stn., Ithaca, NY.

TUMMALA, R. L., HAYNES, D. L. and CROFT, B. A. 1976. Modeling for Pest Management: Concepts, Techniques and Application. Michigan State Univ., E. Lansing.

USDA. 1978. Biological Agents for Pest Control: Status and Prospects. U.S. Dept. Agric., Washington, DC.

USDA. 1980. Basic Principles of Insect Population, Suppression and Management. Agric. Handbook 512. U.S. Dept. Agric., Washington, DC.

USDA. 1980a. Report and Recommendations on Organic Farming. U.S. Dept. Agric., Washington, DC.

USDA. 1981. America's Soil and Water: Condition and Trends. U.S. Dept. Agric., Washington, DC.

USDA. 1981a. Cooperative Gypsy Moth Suppression and Regulatory Program Activities. U.S. Dept. Agric., Forest Service, Washington, DC.

USDA. 1981b. Urban Integrated Pest Management: A Report. U.S. Dept. Agriculture, Science and Education Administration. Extension Committee on Organization and Policy. Univ. of Georgia, Athens.

VAN DER LANN, P. A. 1956. The influence of organic manuring on the development of the potato root eelworm, *Heterodera rostochiensis.* Nematology *1,* 113–125.

WEARING, C. H. 1982. Integrated pest management: Progress and prospects with special reference to horticulture. N. Z. J. Exp. Agric. *10* (1) 87–94.

WELCH, S. M. 1982. Risk-based design of meteorological networks for integrated pest management. *In* Biometeorology in Integrated Pest Management. Academic Press, New York.

WILLIAM, R. D. 1981. Complementary interactions betwen weeds, weed control practices, and pests in horticultural systems. HortScience *16,* 508–513.

10

Future Technology

Over the last two hundred plus years, American agriculture has responded to the various demands and problems thrust upon it. The success of that response owed much to research, and the development and implementation of new technologies. Among the most noteworthy technologies have been chemical aids, such as fertilizers, pesticides, and growth regulators; labor-saving machines, such as tillage equipment, planters, applicators for pesticides and fertilizers, and harvest equipment; and new cultivars designed for maximal yield, pest resistance, and compatibility with existing technologies.

Once again American agriculture is faced with serious demands and problems, some of which result from previously adopted technology. Much of the technology depended on inexpensive energy, but now energy consumes a painful part of the agricultural budget. Chemical aids, because of their energy costs and environmental impact, are no longer the great panacea. In an effort to maintain profits in the face of escalating costs, some farmers neglect the maintenance of their basic resources, leading to diminished soil and water quality.

As these concerns mounted, the new direction for American agriculture became one of sustainability. With awareness came responses, such as those discussed in previous chapters. Technologies associated with problems were not discarded but modified to meet new demands. Chemical pesticides blended with nonchemical pest controls to become a newer technology, integrated pest management. Chemical fertilizers were joined to soil-conditioning methods favored by organic farmers. Energy use was made more efficient and better management of soil and water was started.

The objective of a sustainable agriculture has not yet been achieved; much remains to be accomplished. Most of the future progress in achieving a sustainable agriculture will rely on adoption of newly emerging technology and the creation of even newer forms. Key parts of this new technology will be computer technology, solar technology, and biotechnology. Surveys of agriculture's possible future are available from the Futures Group (1982) and the Battelle Memorial Institute (1983).

Before moving on to a discussion of new technologies for sustaining agriculture, it should be pointed out that substantial research remains yet on existing technologies. These areas were highlighted throughout the preceding chapters. This near-term and future long-term research will require substantial federal funding if a sustainable agriculture is to become a reality. In view of the consequences if we fail, such funding must be given a high priority.

With this need in mind, it is interesting to note that the President's Private Sector Survey on Cost Control has reached some interesting conclusions in its 1983 draft report on the Department of Agriculture. The possible savings have been grouped into various categories, based upon the degree to which they are realistically substantiated, defensible, and supportable on their management merits. The most realistic recommendation, one which is fully substantiated and defensible, is that some $10 billion cost savings are possible for the Department of Agriculture over 3 years. While social, economic, and political factors will influence the implementation of these savings, serious thought should be given to these recommendations as a possible way to fund the nation's future in sustainable agriculture.

Solar Technology

Solar energy, technology and economics being favorable, can presumably be substituted for stationary fuels used on the farm. Such fuels are electricity, natural gas, and LP gas; these energy sources are used on the farm and residence for heating, cooling, drying, lighting, and running various pieces of equipment and appliances. These fuels account for about 22% of farm energy needs.

An average U.S. farm requires about 934 million Btu of energy each year; 84% goes for business operations and 16% for the residential parts of the farm (Table 10.1).

Fuels likely to be replaced by solar energy are electricity and LP gas,

Table 10.1. Farm Use of Stationary Fuels

Activity	Usage by fuel (Btu $\times 10^9$)		
	Electricity	LP gas	Natural gas
Crop drying	—	57904	700
Home heating	2350	29740	36680
Livestock maintenance	21165	5001	—
Poultery breeding	—	19835	700
Water heating	2220	2100	2480
Total (Btu $\times 10^{12}$)	1136.4	119.6	—

Source: Adapted from Heid and Trotter (1982); does not include coal, fuel oil, or wood.

mostly because of costs. With changing inflation and decreasing supplies, it is not easy to predict long-term fuel cost differentials. The present national cost ratio of electricity: LP gas: natural gas per 100,000 Btu is about 3.2:1.4:1. Natural gas presently costs less than solar energy, so it is not likely to be replaced by solar energy unless deregulation and supply problems drive the price of gas upwards. A general discussion on solar energy is presented by Spillman (1981).

Crop Drying

The energetics of crop drying were discussed in Chapter 3. About 90% of the crop-drying energy is used on corn. Other dried crops include rice, hops, hay, peanuts, tobacco, and some fruits and vegetables. The present outlook for solar technology applications to crop drying is quite good. With good designs and multiple-use efficiency, the cost payback period is 3.5–5 years, making solar technology economically feasible (Heid and Trotter 1982).

More research is needed on portable, multiple-use designs. Most of the research in the 1970s produced stationary solar systems designed for drying a specific crop. Systems were usually air-type, flat-plate collectors. Present research is moving toward portable, multiple-use collectors, since ducting to sites and buildings can result in heavy heat loss at 15.2–30.5 m (50–100 ft). However, multiple uses entail variable heat requirements and air flow. Such variability makes it difficult to obtain excellent efficiency under all conditions, thus making some compromise necessary.

Research on solar driers for grain (Fig. 10.1), mostly corn, has produced a number of conclusions. Air-type collectors were shown to be more efficient than liquid types with heat exchangers. In terms of economics and simplicity of on-farm construction, they were deemed better than concentrating high-temperature collectors. Drying of high-moisture grain was not profitable, since stirrers and a back-up system were required. The latter was necessary for periods of low solar radiation and high humidity (Thompson and Pierce 1977; Heid and Aldis 1981). Examples of collector designs in the 1970s are discussed by Heid and Trotter (1982).

More recent examples of solar grain-drying designs are quite promising. One collector, designed both for crop drying and heating of livestock quarters, showed economic competitiveness with conventional energy (Heid and Trotter 1982). Another has multiple uses, portability, flexible airflow, and tiltability. Payback time was estimated at 5.8 years (Heid 1981). Plans are available from the University of Illinois (1979) for a multiple-use, portable collector which could be built for around $1300 to $1400 at today's costs. Interest and acceptance were great enough for the USDA to start a solar grain-drying demonstration project in 1980 (Heid and Trotter 1982).

Other solar designs have been researched for drying of hay, hops, fruits

FIG. 10.1.   Solar grain drier utilizes a flat-plate, air-type collector, as seen on the ground to the left.
Courtesy U.S. Department of Agriculture

and vegetables, as well as the curing of peanuts and tobacco. Again the short period of drying activity makes solar drying uneconomical for commercial applications unless multiple uses are considered.

For example, hay or hop drying is done during the summer. This would leave the solar drier free for use in grain drying and space heating applications. Another possibility is for the sequential use of a drier on tobacco, corn, and then peanuts. Another is for solar energy to be used in a greenhouse for heating of crops and for curing of peanuts and tobacco (McElroy and Krause 1982).

Recent solar-drying research on hay includes that of Bledsoe and Henry (1981) and Morrison and Shove (1981). Drying of hops with solar collectors has been examined by Kranzler (1981), and of vegetables and fruits by Coleman (1981). The curing of peanuts with solar driers was studied by Schlag and Sheppard (1979) and Butler and Troeger (1981), and of tobacco by Cundiff (1981), Henson (1981), and Huang and Toksoy (1981). Guceri (1983) has recently reviewed heated air solar collector use with various crops.

An interesting departure from the heated air-type collectors for drying grain, using a pond as a solar collector, shows promise. The solar pond concept is presently an active research project at the Argonne National Laboratory (1983). The key to the solar pond is salt water. The salt concentration increases with depth, forming gradients that minimize convection currents in the pond. The heavier layers are heated by the sun's rays. Since the layers cannot rise, the heat is trapped. The pond reaches temperatures over 77°C (170°F) in the summer, and remains above 38°C (100°F) in the winter. The temperature peak is reached in September, which is ideal for the grain-drying season. More work is needed on the heat extraction equipment for such a system. The solar pond concept has also been examined at the University of New Mexico, which has one of the oldest projects, at Ohio State University, and in Israel (SERI 1981a). Solar ponds can also be utilized for space and water heating, a topic which we will now examine.

Space and Water Heating

The heating of farm space and water runs a close second to crop drying in terms of solar potential. Some possible uses in this category include the heating of various farm buildings, such as animal quarters and the residence, and the heating of water for these same applications. Solar units for these purposes can be for a single use and still be economical, assuming efficient designs.

The most promising solar uses, in terms of design and ongoing research, are for heating of swine and poultry houses, and providing hot water for dairy operations (Fig. 10.2). This is especially true when solar collectors are used

FIG. 10.2. Solar-heated dairy parlor with a flat-plate roof collector can be used to heat hot water for dairy operations.
Courtesy U.S. Department of Agriculture

both to heat and cool buildings. Under these conditions a payback period of 5 years is likely and was achieved on some 90 solar demonstration livestock farm projects. A number of these designs for swine, poultry, and dairy operations are discussed by Heid and Trotter (1982).

One system can be used year-round in swine houses for farrowing and nursery purposes (Houghton 1980). The solar collector provides both heat and hot water; the payback time appears to be near 5 years. This system is now commercially produced.

Space heating for the farm residence (or many homes for that matter) also appears to be economically feasible, especially in view of energy tax credits. Some systems are already available on the market, and a number of do-it-yourself designs exist. A similar situation also exists for solar hot water systems. Heid and Trotter (1982) discuss both systems. A number of residential solar-heating systems are covered in federal publications (HUD/DOE 1979; SERI 1981a).

Future Uses for Solar Energy

A number of research projects have examined the use of solar energy in several other areas. These have potential applications in agriculture, but do not presently have as great an economic feasibility or promise as the uses already discussed. This may change if energy costs continue to rise. These areas include coolers, distillers, greenhouses, and irrigation. A discussion of these solar energy applications can be found in Heid and Trotter (1982).

A marketing analysis on present and projected solar energy technology was performed by Heid and Trotter (1982), who indicated a promising future for agricultural use, provided that more research and development were conducted to refine the technology. Improved communications and better delivery of the technology to farm users is also a prerequisite. One problem is that agriculture needs low-cost technologies, which take second place to high-cost technologies in terms of marketing and development by the solar industry. Declining federal funding for agricultural applications further complicates the problem. Even so the Department of Energy estimates that by 2000 half of the energy used in agriculture could be replaced by solar energy.

One exception to the above appears to be the marketing and development of photovoltaic systems. As pointed out in Chapter 3, photovoltaic systems presently have some limited use in irrigation systems in the Southwest. Since 8% of all the energy use on the farm is electricity, many other uses for electricity from photovoltaic systems are possible. Besides irrigation, uses in the near future include applications in poultry farming and electric fences. Projections are that solar-produced electricity may become competitive with conventional electricity by the late 1980s (SERI 1981a).

Windpower

Many of the basic practices being adopted to form today's sustainable agriculture were commonplace on farms in the 1800s. One of these was the use of windmills to pump water and generate electricity. The wind is converted to electricity at an efficiency comparable to fossil fuel power plants. A discussion of windpower (Fig. 10.3) is provided by Klueter (1981).

The key to the economic feasibility of windpower in today's agriculture is similar to that with solar energy—multiple uses. Suggested uses of the generated electricity include heating of buildings, crop drying, irrigation, and cooling operations. The economics of wind energy is improved if the system can store electricity for use during windless times and if surplus energy can be sold to utilities through an electrical grid system.

Areas of sufficient wind to provide reasonable energy exist in about one-half of the United States. Very few of the favorable areas exist in the eastern United States. Even in favorable areas, the potential contribution of windmills is thought of as supplementary to solar energy systems. Additional information on wind energy, appropriately enough in view of the foregoing projection, is available from the Solar Energy Research Institute (SERI 1980a), and also from NASA (1982).

FIG. 10.3. This vertical-axis Darrieus-type wind turbine in Texas harvest at least 30% of the energy in wind.

Courtesy U.S. Department of Agriculture

Hydroelectric and Geothermal Power

The use of waterwheels to do work and generate small amounts of electricity is another old practice, once very common in the 1800s. Today's successor, various forms of water turbines, may also increase production of energy on the farm.

Areas in the eastern United States, unable to generate wind energy, are fortunate in that they have many potential hydroelectric sites. Other favorable areas include the Northwest and northcentral states. More than likely, some farms in these areas with favorable hydroelectric sites will derive some of their future energy needs from hydroelectric power.

One particular advantage is that low-head hydrogeneration is constant and controllable, a distinct advantage over wind energy. Sales of surplus energy through an electric grid are probably more realistic because of this advantage. The initial cost and possible legal problems are presently a drawback. A good discussion on hydroelectric power is presented by Matson (1981). Information on construction costs and production expenses is available (DOE 1983).

Geothermal energy can be used to heat buildings and provide energy for other purposes. Some present uses are the heating of homes and buildings in Idaho and Texas, provision of electric power in California, the drying of onions in Nevada, and the raising of fish in Idaho. The major hot ground-water resources for geothermal energy tapping are in the western United States, but some sites do exist in eastern and western coastal areas. Abel and Walker (1981) discuss geothermal energy.

Potential uses of geothermal energy on farms include the heating of greenhouses, homes, animal facilities, and ponds used for aquaculture. A number of agricultural applications are currently in use. For example, a ranch in South Dakota derives heat for buildings, grain drying, and water warming for livestock. Fish farming in California, Colorado, Idaho, Oregon, Utah, and Wyoming take advantage of geothermal resources. Some food processing in Nevada utilizes geothermal energy. Undoubtedly some farms will provide some of their future energy needs through geothermal systems.

Biotechnology

The sustainable agricultural futurist gets especially excited when certain aspects of biotechnology are examined: genetic engineering and biomass. Biotechnology may be the salvation of agriculture someday. The keyword is "someday." Our present knowledge about the molecular biology, biochemistry, genetics, and physiology of plants is not so complete as we would like to think; much more data is needed to assure the successful development and application of biotechnology to agriculture. Commerical applications of biotechnology will probably not be realized until the 1990s. Until then we must depend upon current technologies to sustain agriculture, such as energy conservation and IPM. Still the consensus is favorable for future help from biotechnology.

Genetic Engineering

The manipulation of plants (and animals) to suit agricultural needs has its basis in genetics. The accomplishments of applied genetics during the past several decades have depended on classical genetic methods, that is, conventional breeding and selection for desired characteristics. The successes of this approach are documented in a recent article by Borlaug (1983).

The newer approach to improving plants has its basis in molecular genetics, that is, direct manipulation or engineering on the molecular and cellular levels using techniques such as recombinant DNA, cell fusion, tissue culture, and *in vitro* fertilization. The new genetic engineering should be thought of as a complement or supplement to conventional plant breeding, not as a replacement. A danger exists that the excitement and promise of genetic engineering may lure researchers and funding completely away from conventional plant breeding, which still has much to offer sustainable agriculture (Borlaug 1983). An excellent overview of genetic engineering is available (OTA 1982).

The recombinant DNA approach involves the splicing of different pieces of DNA together and then their introduction into a host. One piece of DNA contains the gene or genes that can express a desirable genetic trait, such as disease resistance or nitrogen fixation. These genes can be selected from any species; the constraints of natural breeding barriers are bypassed. For example, genes from bacteria, other plant species, or even an animal could be introduced into the recipient host, whether it be a plant, animal, or microorganism. The second piece of DNA serves as a carrier or vector. The desirable DNA is attached to the vector, which then transfers the useful, but foreign, DNA into the host's genetic material. Another characteristic of the vector is that its genes are read by the host and then expressed. If done correctly, the genes added to the vector will also be read and expressed. The DNA is introduced into a cell, which must then be regenerated into an organism.

Using this recombinant DNA method, scientists have produced two vaccines that are presently undergoing tests and may appear on the market in the late 1980s. One is a vaccine to prevent a specific strain of hoof and mouth disease in cattle; the other prevents colibacillosis, an intestinal swine infection (McElroy and Krause 1982). Another successful example of the recombinant DNA method is the introduction of antibiotic resistance from a bacterium into a petunia plant (NRC 1984).

Other applications of recombinant DNA, utilizing microorganisms as host cells, also offer promising results for agriculture. These include production of other vaccines, antibiotics, and hormonal growth stimulators. The latter enhance milk and meat production and will be used with cattle, poultry, and swine. Preliminary tests for milk production are good. Microorganisms altered by recombinant DNA may also produce essential amino acids, vitamins, and other additives for use as supplements for animal feeds, or even convert biomass into food for humans and animals. Further off is also the possibility that this approach could be used to produce pesticides from host cells (Miller *et al.* 1983). Many other agricultural uses are also envisioned further in the future (McElroy and Krause 1982). One interesting future application of genetic engineering concerns the issue of diminished germplasm resources. Since molecular engineering has the potential to introduce specific traits from plants outside of compatible breeding populations, it could partially offset losses in genetic diversity. Barton and Brill (1983) provide an interesting discussion on future prospects and the numerous problems impeding progress.

Two problems impose substantial limitations. One is the vector, which must be accepted by the host cell and also be expressed in the host's genetic processes. The most common vector currently used is called the Ti plasmid vector, which is the disease-producing agent of the bacterium *Agrobacterium tumefaciens*. This bacterium causes crown gall disease, that is, plant tumors. When used as a vector, the tumor-causing part of the DNA is removed and replaced by the genes that the researcher wishes to add to the host cell. The main problem is that this vector only works with plants susceptible to crown gall disease. For example, it does not work with such major agricultural crops as corn, rice, and wheat. Viruses could be used as vectors, thus eventually allowing recombinant DNA success with other crops. Much more research is needed, however, with viral vectors.

A second major problem is the regeneration of the host organism from a genetically altered cell. The problem does not occur with single-celled organisms, since regeneration is not needed. Hence we have seen successes with vaccine production using genetically altered bacteria. However, regeneration of a cell into a plant poses considerable difficulty, and into an animal impossible difficulties with our present knowledge. Some plants have been regenerated from cells; these include alfalfa, carrot, petunia, potato, tobacco, and tomato. Others, such as corn, soybeans, and wheat, have not been successfully regenerated.

Some interesting results have been obtained when protoplasts (plant cells with cell wall removed) of sexually incompatible species were fused. This technique, called protoplast fusion, has given rise to somatic hybrid plants, thus offering the promise of new plants not possible through conventional plant breeding. To date this process works only with closely related species in the same family that are sexually incompatible, but not with unrelated species. At this point protoplast fusion offers no control over what genetic information is lost or retained, unlike the recombinant DNA method. Even though some novel plants have resulted, such as a cross between tomato and potato, these plants need additional breeding and selection by conventional means to produce a plant of use in agriculture. Reviews of protoplast fusion, various resulting somatic hybrids, their genetics and problems have been presented by Shepard et al. (1983) and Evans (1983).

Tissue cultures of plants have been moderately successful. Such cultures are useful for the propagation of virus-free plants and the propagation of plants that are either difficult or slow to propagate by conventional methods. Plants processed by commercial tissue culture include asparagus, citrus fruits, orchids, pineapples, and strawberries. However, the difficulty of regenerating whole plants from tissue culture has slowed the use of tissue culture methods with crops such as corn and wheat. Information on the techniques is available in Conger (1981) and Evans et al. (1983).

Of great potential value is the mutagenesis of tissue cultures and selection for mutants showing disease resistance, greater yields, or other desirable qualities. In comparison with the selection of mutants by conventional methods, the tissue culture approach offers savings in space, time, labor, and money. No agriculturally useful mutants have been produced yet. The problems with this approach are discussed by Chaleff (1983).

Tissue cultures often undergo mutation without exposure to mutagens. These spontaneous mutations can involve one or many genes, and some are stable in the progeny of the regenerated plants (Evans and Sharp 1983; NRC 1984). Some stable observed traits for wheat include variations in height, number of side shoots, color, and seed storage proteins. This area may offer some promising developments in the future.

What is the likelihood of successful application of biotechnology to agriculture? The United States is rated as leading the world in the biotechnology race and definite commercial results are expected in agricultural areas (OTA 1984). Vaccines and growth promoters for animal agriculture are predicted to have relatively early commercial success, second only to pharmaceuticals. Success in plant agriculture is not rated among the top three, but is said to hold great promise.

Embryo Transfer

Frozen semen from prize bulls has been used to produce superior dairy cattle. Today, about 70% of dairy cattle are produced by artificial insemination, but only about 3% of beef cattle. Increased production of farm animals by controlled breeding is even more likely with the advent and recent commercialization of frozen embryo transplant technology.

In this method, a cow superior in some traits (e.g., disease resistance or meat quality), is induced with a hormone to produce several times the normal number of eggs. These eggs are fertilized with semen from a superior bull. After 1 week of development, the embryos are removed, stored if necessary, and placed in surrogate mothers. In this way production time for a genetically superior cow is greatly reduced (Battelle Memorial Institute 1983). One cow could easily donate enough embryos to produce 100 superior animals in a few years, instead of the equivalent production time under normal circumstances of 100 plus years.

Further multiplication is possible. The embryo can be surgically split, and then developed into identical twins. Another possibility is the fusion of two embryos, such as from two breeds, and the production of a hybrid animal. Such procedures promise rapid upgrading and genetic improvement of dairy and beef cattle in the near future. Further information on a commercial venture is available (King 1982).

The present limitation is cost; embryo transfers cost about $2000. As costs decrease, or the values of breeding animals rise, dairy cattle will likely be the first farm animals to be bred mostly by embryo transfer. Beef cattle would follow. Eventually, sheep and swine may be bred by embryo transfer.

Biomass

In 1981 the amount of energy derived from biomass, including wastes, for use in the United States was 2.7 quad and is projected to reach 3.5 quad by 1985. The 1981 level was about 3.5% of total U.S. energy consumption. This placed biomass ahead of geothermal, solar, and wind energy combined in terms of energy contribution (Haggin and Krieger 1983). An excellent overview of energy from biomass is available (OTA 1980).

While the energy derived from biomass is impressive, some disquieting questions arise. Is this good or bad for sustainable agriculture and food production? Presently agriculture and forestry utilize about three-fourths of the land in the United States. The amount of surplus land, considering urban and industrial uses, left for increased biomass production appears limited. This suggests diversion of existing land to biomass energy at the expense of food, fiber, or wood products. Wastes or crop residues diverted for biomass may increase erosion and reduce soil productivity, since most crop residues and manures now are returned to the soil. One exception is manure produced on feedlots. Other changes might accompany increased biomass production: altered or reduced wildlife habitats, increased costs for land and food, and a rise in biomass-related injuries (Pimentel *et al.* 1984).

Other problems arise. The diversion of food resources is not necessarily efficient. For example, the amount of corn needed to produce ethanol to fuel a car is nine times that needed to feed one individual. A barrel of ethanol produced in Brazil from sugar cane costs twice as much as a barrel of oil (Pimentel 1982). A little over one-half of the U.S. corn crop would be needed to supply current U.S. gasoline needs with gasohol (Doering III and Peart 1981).

The future use of biomass to produce energy must be considered carefully, so that it does not conflict with the goal of sustainable agriculture. One possible benefit from biomass and wastes for sustainable agriculture is on-site energy production with surplus materials left after food and soil needs are satisfied. We will now examine these possibilities. Extensive information on the production of liquid and gaseous fuels from biomass is available (Wise 1981, 1983a, 1983b).

Alcohol Production

The production of ethanol on the farm through fermentation is possible on a small scale (SERI 1980b, 1981b; Miller 1981). Starting materials can be cereal grains, sugar cane, molasses, sugar beets, wood and wood by-products, cheese whey, potatoes, rice, crop and fruit product residues.

Ethanol produced on the farm is a wet alcohol, since it contains small amounts of water. Production of dry alcohol, which is required in gasohol, requires processing capabilities not likely to exist on farms. However, wet alcohol can be used as a fuel in modified internal combustion engines or by injection into the airstream of a diesel turbocharger as a supplement with diesel fuel.

The use of small amounts of cereal grains, especially if spoiled or of inferior quality, might not be as bad for food production as one might think at first glance. The cereal grain byproduct left after fermentation still retains most of the original nutrients, except for carbohydrates, and could be used in animal feeds and possibly in human foods if the original grain was of high quality.

Crop residues can also be fermented to produce ethanol. The danger here is that most are presently returned to the land for maintenance of soil productivity. If any are used, they should be surplus; if not, some other form of organic matter should be returned to the soil in their place.

Another alcohol might possibly be produced on the farm: methanol. The process is termed pyrolysis, the heating of biomass with little or no air. Gases and liquids are produced, leaving a charcoal residue. The gasification process can be modified to maximize methanol production. Biomass used in this process can be rice hulls, nut shells, wheat straw, and wood wastes. Zerbe (1981) presents a discussion of pyrolysis of biomass.

Methanol can be mixed with gasoline to make gasohol. Some engine modifications are required to burn methanol-derived gasohol, since it is more caustic than ethanol or conventional gasohol. Another drawback is legal restrictions; the Environmental Protection Agency will not allow more than 0.3% by volume of methanol because of emission problems (Haggin and Krieger 1983). Unless this changes, the production of ethanol on the farm is more likely, even considering that the starting biomass for methanol is less deleterious for food production and soil maintenance. Ethanol is also more likely to be purchased by processors for processing into gasohol. The farmer could supply energy indirectly, that is, the income could offset energy costs.

Biogas Production

The pyrolysis process just described can be modified in the gasification process to maximize gas production. The gaseous mixture, termed gasogens, might be used in the future to run stationary engines for irrigation, various farm vehicles, and perhaps boilers. More research is needed for this purpose, since some problems exist (Zerbe 1981).

Methane can be produced from the anaerobic digestion of biomass, including wastes, and even regional resources such as peat (Aiken *et al.* 1983). Information on the production of methane is provided by Stafford *et al.* (1980). Such production might be useful on feedlots, where manure accumulates in large amounts. A few feedlots presently produce methane (SERI 1981b). Methane could be used on farms for heating of buildings and water (Fig. 10.4).

Small-scale methane digestors, suitable for farm needs, are possible. However, the scale limits the supply and the gas quality is different from natural gas. The sale of surplus methane is unlikely, both because of limited supply and quality.

Oils from Seeds

The production of oils from seeds for use as a diesel substitute is an interesting possibility (USDA 1981). More research is needed, however, since the substitute fuel leaves an unacceptable carbon residue in engines. The process (pressing of oil) is relatively simple, and the residue is rich in protein and makes an excellent animal feed. Since a large part of farm equipment is diesel powered, this approach could contribute to energy self-sufficiency. Possible seeds for this purpose could include peanuts, cottonseeds, and sunflowers. Researchers at North Dakota State University are investigating the use of sunflower seed oil. The USDA has also developed a process for converting vegetable oil into diesel fuel, which is presently being engine-tested (USDA 1985).

FIG. 10.4. Manure from swine is used to produce methane for heating purposes at this University of Missouri swine farm. This USDA cooperative project has produced methane continuously since 1976.
Courtesy U.S. Department of Agriculture

Assessment of Alternate Energy

Two possible scenarios exist in regard to farms and alternate forms of energy. Farmers may use alternate forms of energy to approach or even achieve energy self-sufficiency; other farmers may export both energy and food. These are two separate operations that involve different activities and requirements.

The energy exporter would grow certain crops such as trees, corn, or other grains, on a vast scale. These crops would be destined for conversion to alcohols or, less likely, gaseous fuels and probably used for gasohol processing. Corn or grain might possibly be modified through breeding for biomass production. For example, yields of wheat for food are constrained because of milling quality requirements; if quality was not required, as in biomass production, breeders could reportedly increase yields by 30% (Doering III and Peart 1981).

The self-sufficient farm might utilize two or more forms of alternate energy to supplement or possibly replace purchased energy. For example, a mixed crop–livestock farm might use solar energy to heat facilities, provide hot water, and dry grain; produce ethanol to run the pickup truck; and use wind energy to provide some of their electricity.

Large-scale biomass producers may increase the problems of agriculture through competition with food production, degradation of the land through erosion, and competition for water resources. This need not be so if the energy farmers use practices of sustainable agriculture, farm marginal land, and maximize energy production from waste biomass or forest and tree farms. Another interesting possibility is to crop peatlands, swampy areas, and other wetland areas with cattails. The cattail biomass can be converted to gaseous or solid fuels, or used to produce electricity (Aiken et al. 1983). A good discussion on wood as biomass and how biotechnology can greatly increase wood yields is available (Garrett 1981; Farnum et al. 1983). Energy farms may never develop unless they become economically feasible.

Farmers attempting to achieve energy self-sufficiency will not all utilize the same approach. The forms of alternate energy chosen will be influenced by many variables: economics of conventional versus alternative forms of energy, reliability of conventional energy, geographical location, availability of feedstocks for alternate energy, salability of surplus energy, level of management sophistication, legal aspects, and results of present and future research. Because of these factors, predicting what forms will be adopted is difficult at best. However, some generalizations are possible (Doering III and Peart 1981).

Wind and solar energy are likely to increase as suppliers of energy on farms. Some applications exist already, and technology is likely to improve in the future. These improvements, cost reductions as the market expands, and relatively low management requirements will make them attractive options for the farmer. In some instances low-head hydropower will appear attractive, because of its reliability and low management requirements. These systems all are likely to generate surplus electricity for sale to utilities. Farm needs will undoubtedly come first, as the selling price will be substantially lower than the purchase price of energy by the farmer.

Biomass energy production will probably not contribute as large a part of on-farm energy over the next decade as will the preceding forms of energy. Even assuming small-scale operations, some biomass is seasonal, the level of required time and management is higher, the materials are bulky, and handling can be troublesome. Of the biomass options, the ones most likely to be adopted are those that produce liquid rather than gaseous fuels, since the former are more storable, salable, and usable in vehicles or farm machinery.

Assuming technological improvement, the most likely biomass candidate is oil seed processing to prepare diesel fuel substitute. The fuel will be burned directly, surplus could be sold, and the residue utilized for animal feed on the farm or sold. Since about half of the liquid fuel used on farms is diesel, the contribution could be substantial.

Ethanol production is probably a second contender, since the technological level to produce it in the anhydrous form for gasohol is beyond the farmer's managerial level. Wet alcohol production, however, is possible. Some farmers may find it worthwhile to make the necessary equipment modifications in order to utilize the wet alcohol as a fuel. More likely, it will be more attractive to sell it to gasohol processors and use the cash to offset energy needs.

One factor that may change the use of waste biomass for preparation of alcohols and gaseous fuels is the use of newer wastes to maintain soil productivity, thus freeing up more crop residues and manures as feedstocks. Many wastes are being investigated, as discussed in Chapter 7.

Mechanization

The trend toward larger horsepower tractors is (Fig. 10.5) reflected in the five- to sixfold increase in horsepower ratings over the last decade in the United States. Today, U.S. firms are making 650-hp tractors and some foreign manufacturers have introduced a 1000-hp version (McElroy and Krause 1982).

These tractors and their attachments are more efficient in terms of labor

FIG. 10.5.   Improved labor and fuel efficiency are possible with large tractors, such as this 650-hp model. The average horsepower per farm tractor is only 59 today, indicating that many 20- to 25-year-old tractors are still in use. Some of these will undoubtedly be retired and replaced with much larger units.
Courtesy Big Bud Tractors, Inc.

and fuel, and probably more economical over the long run. One example cited by Buckingham (1981) is that of a farmer replacing five tractors and tillage machines with one super-tractor. Fuel consumption was cut in half and one person did the work of six before. With such equipment it is possible to disk 259 ha (640 acres) in one long or two medium-length days.

Such machines are suitable for large farm holdings, such as those in the Great Plains, South, and Pacific Northwest, which grow crops such as corn, cotton, rice, soybeans, and wheat. They are not feasible for use with row crop planters. However, these large tractors can be used with chisel plows, cultivators, disk harrows, and drills, and for laying plastic tile drainage lines. Both conventional and no-till operations can use these large tractors.

Major changes also are occurring in planters, which have increased in width about sixfold over the last decade. Today's planters may be 24 rows wide and equipped with fertilizer attachments. These machines are highly efficient on large acreages. Special forms of planters for minimum tillage are also appearing now. These planters can sow seed and apply fertilizer, herbicide, and insecticide in one pass. Such planters minimize energy needs. Fuel cuts of about 75% are claimed for one minimum tillage planter (McElroy and Krause 1982).

New machinery innovations include a land imprinter (Fig. 10.6) designed for restoration of mismanaged and overgrazed grasslands (USDA 1981). In a trial demonstration, grass yields per acre exceeded controls by 58-fold in 10 months. The machine produces a surface resistant to erosion and runoff, giving the simultaneously prepared and sown seedbed a chance to get established.

Another change is the development of a "knife-in" applicator designed for use with fertilizers and pesticides. This machine applies materials to a depth formerly possible only with a moldboard plow. Such a machine will undoubtedly be of use in no-till operations.

Trends toward larger size and innovative design are also observed with harvesting and on-farm processing equipment. New combines have increased harvesting capacity fivefold over the last 5 years. For example, one new combine boosts corn harvesting capacity to 2000 bu/hr. A new cotton harvester reduces cost by one-sixth and doubles the acreage harvested compared with smaller machines (McElroy and Krause 1982).

Some crops are not suitable for mechanical harvesting because of variation in cultivar size. In time these crops are also going to be mechanically harvested with new types of machines. One example is the pepper, which varies from small chili to big bell types. Profitability dictates that one mechanical harvester should be able to handle all the possible sizes. Such a mechanical harvester for peppers is now a reality (Kovalchuk 1983).

Innovation has also produced a harvester and partial processor for peas.

FIG. 10.6. The land imprinter, developed by Robert Dixon of the USDA, offers great promise for the restoration of overgrazed and mismanaged rangelands. Great success was achieved in a 500-acre trial planting of weeping lovegrass near Fort Huachua, Arizona.

Courtesy U.S. Department of Agriculture

This machine harvests the peas, strips the pods, and separates the peas from the debris in one field operation. Fuel and labor costs are reduced by 10–15% (Becker 1983).

One change likely in the future could affect all field machinery from tractors to harvesters—the concept of controlled traffic (Battelle Memorial Institute 1983). The same paths would be used during each growing season, and would not change from year to year. Equipment, whether used for tillage planting, fertilization, plant protection, or harvesting, would be confined to these paths. This restriction would prevent compaction of the growing area, which causes losses in crop productivity. Significant yield increases have been noted with controlled-traffic farming, except when water is limited. Such a system could also reduce labor if automated guidance systems were utilized.

Another possibility is the incorporation of on-board computers on tractors and planters. Such devices can maximize engine efficiency and save fuel on tractors, when timeliness is not a prime concern. When timeliness is critical, the computer mode can be altered to have the engine produce the maximal work rate. Sensors will monitor field conditions and equipment operation; the feedback will be used to regulate the engine and transmission, thus assuring the desired fuel economy or work rate. Planters will also utilize computers aided by sensors to get uniform crop stands (Battelle Memorial Institute 1983).

Computer innovations that save water and cut costs are already appearing on irrigation equipment. Some of these are laser directed and computer controlled, and use photovoltaic batteries for power (Kanninen 1983). Other recent innovations include computerized neutron field probes to measure root zone moisture, energy-conserving sprinklers, more efficient valves, and better irrigation tubes (Anon. 1983).

A number of factors makes it unlikely that these innovative computer-dependent and larger, more efficient machines will be adopted by all farmers. Some of these factors include the limited acreage of small family farms, topography problems like slopes and field irregularities, roads or bridges unable to handle these large units, plentiful supplies of farm labor, or insufficient cost justification.

Management

The management of agricultural production and processing is undergoing an explosive change. The replacement of family farms by corporate farms has already occurred to a significant extent in some areas of agricultural production, particularly broiler production and cattle feedlots. Hog production is likely to go that way too. Intensive management has made these larger operations more efficient and profitable.

Smaller, as well as large farms, now have a technology that can help them become very efficient in management: the computer. More and more computer applications are being designed for various areas of agriculture, such as bedding plant producers and vegetable growers, computer access information systems, and computer-directed equipment. The promise of this new technology for agriculture has attracted the interest of Congress (Congressional Research Service 1982).

Computer-Assisted Management

Computers are now at a size and price where they can be of assistance to agricultural managers on various levels, from family farms to corporate enterprises. Computers in agriculture can do three basic things: They can provide information on the managerial level about the local operation, which will make it more efficient; they can provide information from external sources to solve problems and improve operations; and they can control certain operations, freeing the individual for other tasks.

Computers already are showing great promise in these applications in the bedding plant industry. If utilized correctly, efficiency and profit can be raised. At this point the technology is geared for greenhouse operations with 25,000–50,000 ft^2 with sales of $250,000–$500,000. It seems likely that in the future computer systems appropriate for smaller operations will be developed (Ball 1983).

The successful adoption of computers in the bedding plant industry is attributed to the fact that appropriate programs and training are available on the commercial level, designed specifically for growers. Applications include cost accounting, crop/space planning, and inventory control. Software is available for these uses. With this software, growers can get a handle on crop profitability, available (but unnoticed) space, workload distribution, shipping and order information, budget planning, and accounting activities. Ball (1983) has a good discussion of these aspects.

A program is also available to determine which of the many crops a grower handles are the most profitable (Aimone 1983). Additional computer programs are available for environmental control and bench movement in the greenhouse (McPhail 1983). A software program to control the greenhouse environment completely and in a manner to produce the best crop at the most economical cost is already under development. Chrysanthemum growers may be able to use this program in their greenhouses as early as 1985 (Anon. 1983a).

A number of computer-accessible information networks are available and of value to agriculture. Some of these will be described briefly, but more details and a complete listing are available elsewhere (Congressional Research

Service 1982). About two dozen plus networks are in use, and some 15 are in the experimental stages. The USDA (1985) is now actively involved with computer models and systems that will benefit the farmer in the near future.

One well-known network deals in information retrieval, giving bibliographic citations and even abstracts from agricultural publications. This network is called AGRICOLA (Agricultural Online Access) and is sponsored by the U.S. Department of Agriculture and the National Agricultural Library. I might add that it was quite valuable in providing information sources for this book.

Others deal in problem solving, such as the nationwide OSU (Oklahoma State University) Farm Management Programs. This network specifically solves problems for farm management. Others deal in the marketing of specific commodities, such as the ECI (Egg Clearinghouse, Inc.) and the EMA (Electronic Marketing Association) for lamb marketing networks.

Another provides the latest marketing information, such as weather conditions, market conditions, and analyses and recommendations on future prices (INSTANT UPDATE). AGNET is another similar information network. Both can supply information via personal computers and cable TV systems. One even deals with the latest research in agriculture and forestry supported by public funds (CRIS, Current Research Information System). Computer networks can be more specific and regional, such as those utilized to schedule irrigation in Wisconsin and California. The former is sponsored by the University of Wisconsin Extension Service, while the latter is avilable through a private management service (Curwen 1984; Ferguson 1984).

Some of the experimental information networks have much to offer and may be converted to full operational status. Some deal with the provision of technical assistance, how-to information, and education for farmers. Others deal with pesticides, integrated pest management, cattle trading, financial aspects, and computer training.

Computer Constraints

Computer use on the farm, especially the smaller units or microprocessors, will undoubtedly increase in the future. Some large corporate farms are using computers now, since they can justify them as cost effective. Estimates by the Kellogg Foundation are that by 1990 some 75% of the commercial farms and 90% of the county extension offices will be utilizing computers. Whether this will become reality depends much upon the resolving of current problems and issues.

The area of computer technology for agriculture is not without problems and issues of controversial nature, as was indicated in depth by the Congressional Research Service (1982). It is beyond the scope of this discussion to explore all of these, and only a few will be mentioned.

One problem is perhaps best called "computer anxiety," fear of computers exhibited by some people. The best way to resolve this is through education and simplification of programming. Those presently in farming are reached best through workshops held by county extension offices. Future farmers should be required to be "computer literate" when they graduate from land grant universities. This direction is an imperative if the potential of computers in agriculture is to be realized. Essentially, a generation of farm managers having computer literacy is needed before the full potential of computers can be realized in agriculture.

In some cases private enterprise may provide the training. Since this is a profit venture, the training is usually attached to the purchase of a computer. One such package, which embodies a high degree of integrity, is that provided by Summersun Technical Service (Ball 1983). Besides training, a followup, toll-free telephone help line is available, as problems develop during later use. This package is for greenhouse management. Because such private operations seek profit, the small farmer or low-cost communication user might not be a suitable market.

Another disturbing aspect is the present confusion about information technology. A bewildering array of computers exists, but which ones are best for the farmer? Where is software designed for the farmer available? Some software exists for the farmer, but much more development is needed. What information networks for agriculture are available? Who can help the farmer? Who will point out the advantages and cost effectiveness? Who will reach the small farmer with information on this new technology? Who will make the instrumentation and information accessibility available to smaller farmers, including some subsidies if necessary? Will copyright cause problems?

Answers to a few of these questions are now becoming available. For example, at no cost farmers can find answers to three questions. Can my operation benefit from computerization? What microcomputer will suit my needs best? What programs will help me to improve management, record keeping, and bookkeeping? These answers are provided through Apple, IBM, and Radio Shack dealers. The farmer enters information and the AgDisk Demonstration and Evaluator Program gives the answers (Anon. 1984).

The other questions will be answered as the federal and state governments and private enterprise seek out and finalize their roles. Undoubtedly there will be some sharing, such as in information services. Perhaps standardization of equipment will occur. Hopefully, accessibility to computer technology will be available for even small farming units. A big plus would be the creation of a central computer facility, which would serve as a library, clearing house, and distribution center.

Only time and awareness will resolve these issues. As this new technology emerges, Congress is examining its role in agriculture (Congressional Research Service 1983). Information technology will play some role in making agriculture more sustainable through improvement of management efficiency.

References

ABEL, F. and WALKER, B. 1981. Tapping geothermal energy, heat from within the earth. *In* Cutting Energy Costs (The 1980 Yearbook of Agriculture). J. Hayes (Editor). U.S. Dept. Agric., Washington, DC.

AIKEN, R. G., HEICHEL, G. H., FARNHAM, R. S., PRATT, D. C., SCHERTZ, C. E. and SCHULER, R. T. 1983. Analysis of energy inputs for peat and peatland biomass development. Minn. Agric. Exp. Stn. Tech. Bull. *AD-TB-2209.*

AIMONE, T. 1983. Ranking crop profitability. Grower Talks *47* (1) 52–55.

ANON. 1983. Items on irrigation. Amer. Fruit Grower *103* (4) 24.

ANON. 1983a. Computerized crops. Amer. Veg. Grower *31* (5) 42–43.

ANON. 1984. Free computer help. American Fruit Grower *104* (3) 69.

ARGONNE NATIONAL LABORATORY. 1983. Argonne Solar Pond Gets State Funding. News from Argonne National Laboratory, Office of Public Affairs. No. 83–57 (August 2).

BALL, V. 1983. Computers for growers are here. Grower Talks *47* (1) 20–27.

BARTON, K. A. and BRILL, W. J. 1983. Prospects in plant genetic engineering Science *219,* 671–676.

BATTELLE MEMORIAL INSTITUTE. 1983. Agriculture 2000: A Look at the Future. Battelle Press, Columbus, OH.

BECKER, R. F. 1983. Mechanical pod strippers. Amer. Veg. Grower *31* (6) 14.

BLEDSOE, B. L. and HENRY, Z. A. 1981. Solar drying of large round bales of high moisture hay. *In* Agricultural Energy, Vol. 1. ASAE Publ. 3–81. Amer. Soc. Agric. Eng., St. Joseph, MI.

BORLAUG, N. E. 1983. Contributions of conventional plant breeding to food production. Science *219,* 689–693.

BUCKINGHAM, F. 1981. Big new hitch for big rigs. Farm Show *5* (2) 15.

BUTLER, J. L. and TROEGER, J. M. 1981. Drying peanuts using solar energy stored in a rock bed. *In* Agricultural Energy, Vol. 1. ASAE Publ. 3–81. Amer. Soc. Agric. Eng., St. Joseph, MI.

CHALEFF, R. S. 1983. Isolation of agronomically useful mutants from plant cell cultures. Science *219,* 676–682.

COLEMAN, R. E. 1981. Building a Low-Cost Solar Dryer Incorporating a Concentrating Reflector. U.S. Citrus and Subtropical Products Laboratory, Agric. Res. Serv., Winter Haven, FL.

CONGER, B. V. 1981. Cloning Agricultural Plants Via in Vitro Techniques. CRC Press, Boca Raton, FL.

CONGRESSIONAL RESEARCH SERVICE. 1983. Information Technology for Agricultural America. Report printed for the Committee on Agriculture, U.S. House of Representatives. Superintendent of Documents, Washington, DC.

CUNDIFF, J. S. 1981. A renewable resource system for curing tobacco. *In* Agricultural Energy, Vol. 1. ASAE Publ. 3–81. Amer. Soc. Agric. Eng., St. Joseph, MI.

CURWEN, D. 1984. Schedule irrigation with WISP. Amer. Veg. Grower *32* (4) 6–7.

DOE 1983. Hydroelectric Plant Construction Cost and Annual Production Expenses—1980. Department of Energy, Washington, DC.

DOERING III, O. C. and PEART, R. M. 1981. How much extra energy can farms produce. *In* Cutting Energy Costs (The 1980 Yearbook of Agriculture). J. Hayes (Editor). U.S. Dept. Agric., Washington, DC.

EVANS, D. A. 1983. Agricultural applications of plant protoplast fusion. Bio/Technology *1* (3) 253–261.

EVANS, D. A. and SHARP, W. R. 1983. Single gene mutations in tomato plants regenerated from tissue culture. Science *221,* 949–951.

EVANS, D., SHARP, W. R., AMMIRATO, P. and YAMADA, Y. 1983. Handbook of Plant Cell Culture: Techniques for Propagation and Breeding. Vol. I. Macmillan, New York.

FARNUM, P., TIMMIS, R. and KULP, J. L.   1983.   Biotechnology of forest yield. Science *219*, 694–702.

FERGUSON, J.   1984.   New firm fine-tunes growers' irrigation. Amer. Veg. Grower *32* (4) 8.

FUTURES GROUP.   1982.   Impending Agricultural Technologies: Insights from Leading Scientists. Vols. 1 and 2. The Futures Group, Glastonbury, CT.

GARRETT, L. D.   1981.   Forests and woodlands—stored energy for our use. *In* Cutting Energy Costs (The 1980 Yearbook of Agriculture). J. Hayes (Editor). U.S. Dept. Agric., Washington, DC.

GUCERI, S. I.   1983.   Solar dehydration. *In* Sustainable Food Systems. D. Knorr (Editor). AVI Publishing Co., Westport, CT.

HAGGIN, J. and KRIEGER, J. H.   1983.   Biomass becoming more important in U.S. energy mix. Chem. Eng. News (March 14), 28–30.

HEID, JR., W. G.   1981.   An Evaluation of the Young Flat-Plate Solar Collector for Multiple Farm Uses. Agric. Econ. Report No. 466. U.S. Dept. of Agric., Washington, DC.

HEID, JR., W. G. and ALDIS, D. F.   1981.   Solar-Supplemented Natural Air Drying of Shelled Corn. Tech. Bull. No. 1654. Econ. and Stat. Service, U.S. Dept. Agric., Washington, DC.

HEID, JR., W. G. and TROTTER, W. K.   1982.   Progress of Solar Technology and Potential Farm Uses. Agric. Econ. Rept. No. 489. U.S. Dept. Agric., Washington, DC.

HENSON, W. W.   1981.   Energy for curing burley tobacco. *In* Agricultural Energy, Vol. 1. ASAE Publ. 3–81. Amer. Soc. Agric. Eng., St. Joseph, MI.

HOUGHTON, D.   1980.   Solar goes sophisticated. *In* Successful Farming. Meredith Pub. Co., Des Moines, IA.

HUANG, B. K. and TOKSOY, M.   1981.   Greenhouse solar system for effective year-round solar energy utilization in agricultural production. *In* Agricultural Energy, Vol. 1. ASAE Publ. 3–81. Amer. Soc. Agric. Eng., St. Joseph, MI.

HUD/DOE.   1979.   A Survey of Passive Solar Buildings. U.S. Dept. of Housing and Urban Development and U.S. Dept. of Energy, Washington, DC.

KANNINEN, E.   1983.   Apply water where and when it's needed. Amer. Veg. Grower *31* (6) 17–20.

KING, W.   1982.   Microscopic techniques have a gigantic effect on cattle breeding industry. New York Times (December 6), A–16.

KLUETER, H. H.   1981.   Windpower can save your energy dollars. *In* Cutting Energy Costs (The 1980 Yearbook of Agriculture). J. Hayes (Editor). U.S. Dept. of Agric., Washington, DC.

KOVALCHUK, S.   1983.   A mechanical pepper harvester. Amer. Veg. Grower *31* (6) 8–9, 22.

KRANZLER, G. A.   1981.   Solar hop drying. *In* Agricultural Energy, Vol. 1. ASAE Publ. 3–81. Amer. Soc. Agric. Eng., St. Joseph, MI.

MATSON, W. E.   1981.   How we can double hydroelectric power. *In* Cutting Energy Costs (The 1980 Yearbook of Agriculture). J. Hayes (Editor). U.S. Dept. Agric., Washington, DC.

McELROY, R. G. and KRAUSE, K. R.   1982.   New Technologies to Raise Agricultural Efficiencies. Agric. Inf. Bull. No. 453. U.S. Dept. of Agric., Washington, DC.

McPHAIL, L.   1983.   Motion control in Canada. Oglevee computers—masters of environment. Grower Talks *47* (1) 36–40, 56–59.

MILLER, D. L.   1981.   The ABC's of making farm alcohol and gas. *In* Cutting Energy Costs (The 1980 Yearbook of Agriculture). J. Hayes (Editor). U.S. Dept. Agric., Washington, DC.

MILLER, L. K., LINGG, A. J. and BULLA, JR., L. A.   1983.   Bacterial, viral and fungal insecticides. Science *219* 715–721.

MORRISON, D. W. and SHOVE, G. C.   1981.   Solar drying of large round hay bales. *In* Agricultural Energy, Vol. 1. ASAE Publ. 3–81. Amer. Soc. Agric. Eng., St. Joseph, MI.

NASA.   1982.   Capturing Energy from the Wind. National Aeronautics and Space Admin., Washington, DC.

NRC.   1984.   Genetic Engineering of Plants: Agricultural Research Opportunities and Policy Concerns. National Academy Press, Washington, DC.

OTA. 1980. Energy from Biological Processes. Office of Technology Assessment, Washington, DC.

OTA. 1982. Genetic Technology: A New Frontier. Office of Technology Assessment. Westview Press, Boulder, CO.

OTA. 1984. Commercial Biotechnology, An International Analysis. Office of Technology Assessment, Washington, DC.

PIMENTEL, D. 1982. Biomass energy. BioScience *32,*769.

PIMENTEL, D., FRIED, C., OLSON, L., SCHMIDT, S., WAGNER-JOHNSON, K., WESTMAN, A., WHELAN, A., FOGLIA, K., POOLE, P., KLEIN, T., SOBIN, R. and BOCHNER, A. 1984. Environmental and social costs of biomass energy. BioScience *34,* 89–94.

SCHLAG, J. H. and SHEPPARD, A. P. 1979. Demonstration of Collectors and Instrumentation for Application to Solar Drying of Peanuts, Tobacco and Forage. Final Report of School of Electrical Engineering, Georgia Institute of Technology, Atlanta. Prepared for the Dept. of Energy, Washington, DC.

SERI. 1980a. Wind Energy Information Directory. Solar Energy Research Institute, Washington, DC.

SERI. 1980b. Fuel from Farms: A Guide to Small-Scale Ethanol Production. Solar Energy Research Institute, Washington, DC.

SERI. 1981a. A Guidebook to Renewable Energy Technologies. Solar Energy Research Institute, Washington, DC.

SERI. 1981b. Fermentation Guide for Common Grains: A Step-by-Step Procedure for Small-Scale Ethanol Fuel Production. Solar Energy Research Institute, Washington, DC.

SHEPARD, J. F., BIDNEY, D., BARSBY, T. and KEMBLE, R. 1983. Genetic transfer in plants through interspecific protoplast fusion. Science *219* 683–688.

SPILLMAN, C. K. 1981. Capturing and storing energy from the sun. *In* Cutting Energy Costs (The 1980 Yearbook of Agriculture). J. Hayes (Editor). U.S. Dept. Agric., Washington, DC.

STAFFORD, D. A., HAWKES, D. L. and HORTON, H. R. 1980. Methane Production from Waste Organic Matter. CRC Press, Boca Raton, FL.

THOMPSON, T. L. and PIERCE, R. D. 1977. Where does solar grain drying fit. *In* Proc. 1977 Solar Grain Drying Conference, Univ. of Illinois, Urbana–Champaign.

UNIVERSITY OF ILLINOIS. 1979. Illionis Plan No. SP-546 Portable Solar Collector. Dep. Agric. Eng., Urbana–Champaign, IL 61801. ($1.)

USDA. 1985. Research Progress in 1984, U.S. Dept. of Agric. Washington, DC.

USDA. 1981. Sunflower Oil May be Farmers' Own Source of Diesel Fuel. Land Imprinter Does Job of Restoring Rangelands. U.S. Dept. Agric., S & E Newsmakers (Science and Education), Sept. (4) 17, 20.

WISE, D. L. 1981. Fuel Gas Production from Biomass. Vols. 1 and 2. CRC Press, Boca Raton, FL.

WISE, D. L. 1983a. Liquid Fuel Developments. CRC Press, Boca Raton, FL.

WISE, D. L. 1983b. Fuel Gas Systems. CRC Press, Boca Raton, FL.

ZERBE, J. I. 1981. Turning farm wastes into usable energy. *In* Cutting Energy Costs (The 1980 Yearbook of Agriculture). J. Hayes (Editor). U.S. Dept. Agric., Washington, DC.

Index

235

Related AVI Books

Breeding Field Crops, 3rd Edition
   Poehlman
Breeding Vegetable Crops
   Bassett
Commercial Chicken Production Manual, 3rd Edition
   North
Fish Farming Handbook
   Brown and Gratzek
Food, Nutrition and Health
   Clydesdale and Francis
Fundamentals of Entomology and Plant Pathology, 2nd Edition
   Pyenson
Introduction to Freshwater Vegetation
   Riemer
Introduction to Plant and Diseases: Identification and Management
   Lucas, Campbell, and Lucas
Plant Health Handbook
   Pyenson
Plant Physiology in Relation to Horticulture, 2nd Edition
   Bleasdale
Postharvest: An Introduction to the Physiology and Handling of
 Fruit and Vegetables
   Wills, Lee, Graham, McGlasson, and Hall
Sustainable Food Systems
   Knorr
Swine Production and Nutrition
   Pond and Maner
Vegetable Growing Handbook, 2nd Edition
   Splittstoesser
World Fish Farming: Cultivation and Economics, 2nd Edition
   Brown
World Vegetables: Principles, Production and Nutritive Values
   Yamaguchi